The first detail structure map (1922) of the famous Ira anticline in Scurry and Mitchell counties led to the first West Texas oil discovery in Mitchell county and, eventually, to the finding of the mammoth Sacroc fields in Scurry County. The nosing of the anticline into Mitchell county and across the Colorado river was frequently referred to as Hardrock (surface) by Walter Lechner's friend, geologist Leon E. English, who did the field work for Foster and Reiter, which was responsible for Loutex Oil Company going into Scurry County. The small town of Ira is shown in the box in the north center of the anticline. (W.W. Lechner files)

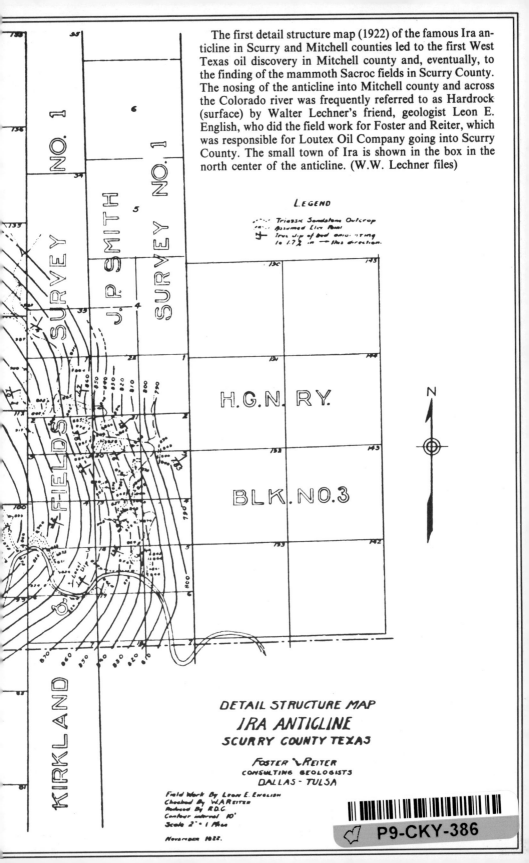

LEGEND

Triassic Sandstone Outcrop
Assumed Live Pool
True dip of bed amounting to 1.7½ in — this direction.

N

DETAIL STRUCTURE MAP
IRA ANTICLINE
SCURRY COUNTY TEXAS

FOSTER & REITER
CONSULTING GEOLOGISTS
DALLAS - TULSA

Field Work By Leon E. English
Checked By W.A.Reiter
Reduced By R.D.C.
Contour interval 10'
Scale 2" = 1 Mile

November 1922.

An Oilman's Oilman

An Oilman's Oilman

A Biographical Treatment of Walter W. Lechner

By James A. Clark

*Editor's Note: Special credit is given to
BARBARA SCHUESSLER,
who assisted Mr. Clark in preparing
this book and who skillfully directed
the book's completion after Mr. Clark's death.*

Edited by Judith King

Gulf Publishing Company · Book Division
Houston, Texas

An Oilman's Oilman

A Biographical Treatment of Walter W. Lechner

Library of Congress
Catalog Card Number:
75-5318

ISBN: 0-88415-633-8

This Book Is Dedicated to
Ruth Nowlin Lechner
whose great love for and stabilizing influence
on Walter have been responsible for the
success of the man this book is about.

Acknowledgments

Much of this book is based on research done for *The Last Boom* (Random House, 1972, Clark and Halbouty), a book about the great East Texas oil field. It is also true that much of *The Last Boom* is based on research for this book.

Most of the information has come from the volumes of diaries kept by Walter Lechner as well as countless interviews with him and photographs he provided. Both Barney Skipper and his wife, Mary, supplied valuable background material, especially for Chapter 7.

Other individuals who contributed considerably to the research for this book include Captain M.T. "Lone Wolf" Gonzaullas, Angus Wynne, John Wagner, W.I. Nowlin, R.L. Foree, and A.D. Moore. Still others are Bill Harrison, E.J. Moran, Paul Barr, The Hon. Allan Shivers and President Lyndon Johnson.

Documents and other sources of information came from *The Handbook of Texas, The Encyclopedia of American Facts and Dates,* 6th Edition, 1972, Crowell; *The East Texas Oil Field, 1930-1950,* by the East Texas Engineering Association, 1953; *The Gregg County 35th Anniversary of the East Texas Field* booklet; *The Longview News; The Wichita Falls Times; The Snyder News* and *The Oil Weekly* (now *Oil World*).

Other sources included *The Austin American-Statesman, The Corsicana Sun,* the Sam Rayburn Library, Gregg County records, Mickey Herskowitz, *The Dallas News, The Fort Worth Star-Telegram,* the Scurry County Historical Society, files of the Energy Research and Education Foundation at Rice University and files of the Yount-Lee Oil Company.

Foreword

This book, which largely traces the important aspects of the life of Walter W. Lechner of Dallas, is about a man who is typical of those responsible for the progress and prosperity in Texas and to a great extent for the prosperity of America and all its people since the birth of the liquid fuel age at Spindletop, near Beaumont, Texas, in 1901.

Walter Lechner and his fellow independent oil explorers and producers have found almost 90 percent of all oil found in this country since 1859. This book tells something of such a man.

The book is titled "An Oilman's Oilman" and I know of no title that could better describe Walter. His whole life has been devoted to finding oil. He has never been identified as one of the "big rich," but his presence and influence have been felt in his city, his county, his state, and his nation. He has lived a simple, decent, useful life, most of it in common with a beautiful and intelligent wife who typifies women of Texas since the beginning of the Republic.

I am proud to call Walter my friend. I once appointed him to the office of chairman of the Texas Game, Fish and Oyster Commission (later Game and Fish Commission, now Parks and Wildlife). He has always been a true conservationist with a deep interest in parks and recreation. He served his state with distinction.

ALLAN SHIVERS

Contents

Introduction

The more I knew about Walter Lechner, the more I wanted to write a book about him. He is the kind of oilman I think all oilmen should be. He has been a marked success in his business, but he hasn't let it go to his head. Contrary to many who have had his success, he doesn't sound off every day about the state of the nation. He doesn't tell his doctor how to operate on him. He hasn't bought hotels or newspapers or produced motion pictures. His name is seldom in the gossip columns, despite his highly active role in state and city affairs when called upon. He lives simply in an apartment with his wife who is as reserved and quiet as he is and who has been his guide to success and happiness.

Yet Walter Lechner is one of the best independent explorers for and producers of oil and gas in the country. His track record of discoveries is remarkably good. He has worked in almost every facet of production, starting as a roustabout on an old salt dome rotary rig. He has made good deals and bad ones, but he has never shunned hard work. So I believe he is truly an oilman's oilman. If there were more like him, the industry's bad image might not exist.

I believe there is not only much for the oil industry to learn in this book but something the average lay person should know about most good oilmen. They are not flamboyant, lavish spenders and shouters. They don't all live in mansions. All of them do not commit the normal sins of the newly rich. And certainly all of them are not wheeler-dealers. Most of them are hard-working, decent, talented men, devoted husbands and family men in the truest American mold.

Walter, therefore, is not an unusual man. He typifies those who search for, find, produce and dispose of oil and gas. He is one of those who has helped make his fellow man more prosperous and contributed in real measure to the progress of his city, his state and his nation.

It's good to know there are people such as Walter and Ruth Lechner.

JAMES A. CLARK

Chapter 1

It was to be a curious field trip. Only a few months before, Walter Lechner had returned from the battlegrounds of World War I. Now in Dallas, with not much to do, he and his father had been invited by a friend to spend a few days in Harrison County, Texas.

The friend, Dr. Hugh H. Tucker, was a hard-rock geologist who planned to do some detective work around the creeks and streams and woods east of Marshall in extreme East Texas. He wanted company. Walter and his father went along for the fresh air while Dr. Tucker searched for something—without explaining what.

Early the first day, the doctor stopped to examine what seemed to Walter to be an outcropping of coal on a creek bank. He chipped off a sample and turned it over several times, studying the stratification and other features. Then he moved to a higher elevation and surveyed the area as far as he could see. He had been there many times before and the geology of the area fascinated him.

At last, Dr. Tucker turned to the Lechners and predicted that some day a great reservoir of relatively shallow oil would be found about 45 miles west of there over in Gregg County, around or beyond Longview. He volunteered no more about his theory or the reasons for his belief. Throughout his life Walter Lechner would never find another geologist who could explain how Dr. Tucker arrived at his conclusion—based on the scant evidence of an outcrop of what seemed to be coal in the eastward edge of East Texas.

Walter, then not quite 30, was never able to dismiss the words of Dr. Tucker from his mind. He had worked at various jobs, had fought a war in France and by now had made up his mind to be an oilman for the rest of his life.

1

Ten or eleven years later when Walter Lechner finally returned to the Tucker idea, it led him to the pot of gold at the end of the rainbow which true wildcatters always seek and only a few ever find.

Oil provided a golden era for this country—a century which extended from 1859 until about 1959.

During that period, the United States emerged from a woodshed technology to the world's most progressive and prosperous industrial nation. Due largely to the bounty of oil, the American standard of living skyrocketed. The greatest dreams and hopes of many people in the history of mankind were realized. The United States became in almost every respect the most powerful nation on earth.

The discovery of oil in Pennsylvania in 1859 established a new and vigorous, vital element of life that brought better and cheaper illumination and ample lubrication to grease the machines of the expanding industrial age and to provide the muscle that helped the United States win the Civil War.

While coal and kerosene were still the major sources of light and power, Spindletop provided evidence, just as the twentieth century began, that there was sufficient oil to supply an age of liquid fuel. Oil made possible the full potential of the industrial age, including mass production and the elimination of child labor and the sweatshop.

Liquid fuel in the form of petroleum also gave birth to the age of mobility. It made the motor vehicle and the airplane possible. It provided cleaner, faster, more efficient, more comfortable and more profitable water transportation when it replaced coal, which had replaced the wind and the slave galley. Liquid fuel brought the highways that linked people together, creating a more cohesive society and a more powerful nation than the world had ever known.

Petroleum provided the incentive for the development of a breed of man the world has seldom seen. This type of man spanned all levels of the educational, cultural and social complex. It was a breed born of a marriage of individual freedom and real opportunity with which these men could exercise their minds, resourcefulness, ingenuity and daring. Petroleum also put them in the arena of nature's constant challenge.

When Walter Lechner was born in 1890, the United States was in second place, behind Russia, as an oil-producing nation. The days of the great oil booms in Pennsylvania, West Virginia, Ohio, Indiana and New York had passed and oilmen said there would be no more booms. At that time, in fact, no one considered oil to be a fuel of great importance and did not anticipate that it ever would be.

The accidental discovery of oil in Corsicana, the curtain raiser of the industry in Texas, was still a year or two away. No one considered Texas a state with a very bright oil future.

Walter Lechner was one of thousands of oilmen born in the years spanning the nineteenth and twentieth centuries. The oil industry was still in the tenacious grasp of the "octopus" known as Standard Oil and would be so dominated until late in the twentieth year of his life. Then new integrated companies, such as Gulf, Texaco, Humble and Sun Oil, shot up out of the gushers of Spindletop in Texas where Standard was persona non grata under new, but vigorously enforced, antitrust laws.

Walter was born into the era of such giants as John Davison Rockefeller, the clerk who became the king of oil; Samuel Langhorne Clemens, who reached literary immortality as Mark Twain; William Randolph Hearst, the titan of yellow journalism; and William Jennings Bryan, the golden-voiced defender of Protestant fundamentalism, to mention only a few of a host of legendary figures.

Benjamin Harrison, a Republican, was President, the twenty-third of the nation. The governor of Texas was the illustrious Sul Ross, a native of Iowa. The state was still one election away from its first native governor, James Stephen Hogg of Rusk.

The year Walter was born, Texas produced less than 500 barrels of oil worth an average of 77 cents a barrel. Pennsylvania produced far more than half the nation's oil and the bulk of the remainder came from Ohio. The value of all oil produced in the country for 1890 was slightly over $35 million.

In all America there were slightly more than 27,000 wells and in that year only 742 were drilled. Furthermore, drilling was declining at an alarming rate. Petroleum, as an industry, seemed to be about finished in America.

All of this was true in 1890, but the country was on the edge of magnificent breakthroughs. The manufacture of Charles E. Duryea's first horseless carriage was only two years away. The Wright brothers' flight at Kitty Hawk would come in 12 years. And Henry Ford's Model T was just 18 years in the future. Of even more significance, his own Texas was destined to give birth to the liquid fuel age before Walter Lechner's twelfth birthday. It would happen about 250 miles south of Kaufman County, near Beaumont, with the discovery of the fabulous Lucas Gusher in the field that became known around the world as Spindletop.

So, October 13, 1890, was in many respects a favorable day for a future Texas oilman to be born . . . one who started life as the son of resourceful, intelligent, ambitious parents, endowed with character and integrity.

The Lechners were a kind of composite Texas family, drawn to this vast and promising land as American life flowed west. His father Philip was a native of Ohio and his mother Priscilla—born Mary Priscilla Truss in Verona, Mississippi—grew up in Tupelo. Priscilla was descended from the rugged clan of Scotch-Irish who found the deep South to their liking. Her family moved to Texas from Mississippi by ox train about the time the Andrew Lechners were moving into the state.

Walter's paternal grandfather, Andrew Lechner, was born in 1819 in Bavaria, the gay and gentle southland of Germany. The Bavarians have been known throughout history as light-hearted folk, fond of malt and music, slow to anger and quick to dream.

Andrew Lechner married in 1845 and brought his bride to America three years later on April 18, 1848. Their ship had sailed the Atlantic for 40 days and nights before making port in New York. There they joined the grand adventure that was to make this country the melting pot of the world. Andrew became a United States citizen in the shortest possible time after his arrival. He was naturalized in the County Court of Common Pleas in Hamilton County, Ohio, on October 11, 1852.

The young German couple later moved to Cincinnati, where the next five years passed quietly. Their next move was to Harrison in Hamilton County, Ohio, where Philip was born in 1857.

As the Civil War was winding down in 1865, the Lechners resettled in Grayville, Illinois. The family remained there until 1877 when Andrew pulled up stakes and set out for Williamson

County, Texas, to spend a year. Then he made what would be his last move to the town of Valley View—one of five settlements of the same name in Texas—a short distance from Colquitt.

What had brought them irresistibly to this place was that great magnet of the late nineteenth century—cheap land. In Kaufman County, the promise of the good earth attracted several Bavarian families—the Bitners, the Snyders and the Lechners. They established a small German settlement and industriously worked the land.

There Andrew Lechner spent the last 16 years of his full and mobile life. Death came almost instantly, a few days before his seventy-eighth birthday. Until that moment he had never been ill. He coughed two or three times. A blood vessel burst in his throat. And in a matter of minutes, before help could be provided, Andrew Lechner was dead. The date was March 15, 1897, when Walter was six.

When he moved to Texas with his family, Philip became a Baptist and indirectly through the church met Priscilla. Strangely, both Philip and Priscilla had blind brothers who attended a church school for the blind in Austin. While visiting his brother at the school, Philip became a friend of Priscilla's brother. Once Priscilla's brother invited Philip to his home in Mesquite, east of Dallas, and there he met Priscilla. On November 2, 1884, Philip and Priscilla were married in Mesquite by Priscilla's grandfather, a Baptist minister.

The Philip Lechners were a solid, close and loving family. Philip was a deacon in the church in Terrell. Priscilla saw that Walter and his brother and sisters never missed Sunday school or church.

In 1890, when Walter William Lechner was born, the predominant political issue across the country, as well as in Texas, was the rising specter of monopoly which brought about the Sherman Antitrust Act. In Texas, another major issue was populism, which for several years challenged the Democratic Party. Nineteenth-century wage earners received less than one-fifth of what 1973 workers earned, but in that early period the cost of living was relative. Walter was born in a rural county where both costs and income were even lower.

His father owned a small farm two miles from town, which he rented to a tenant named Tom Griffiths, and he operated a grocery store in Terrell in partnership with Brewer Hale. This was picture window Texas with rolling plains, searing summers, cold winters and vast expanses of cattle and farmlands.

Walter was the first Lechner born in Terrell after his parents moved there from Colquitt. He was the first boy in the family. Minnie Gertrude, May Ola and Martha Grace were born in Colquitt, a small dot on the map of Texas, west of Terrell in Kaufman County. Martha Grace, born the year before Walter, died in infancy. Her death influenced the family's decision to move to Terrell where a physician was available. Mary Ethel was born in 1894. Walter was eight and a half years old when his only brother, Philip Andrew Lechner, Jr., came along and he was more than 12 years old when his baby sister, Lucille Elizabeth, was born.

In any contest to pick an average, small-town Texas farmboy, Walter Lechner would certainly have reached the finals. He was named for an uncle, William Henry. The "Walter" was tacked on to give the boy a sense of his own identity. Before his teens, he divided his time working after school in Philip Lechner's grocery and working in the summer and on weekends for the man who rented his father's farm. He chopped and picked cotton, worked in the hay fields, milked cows and did the usual chores to earn money. It was hard work but invigorating for both mind and character.

In the Terrell elementary school, Walter was a good student. He dreamed of becoming a locomotive engineer. The railroad was a romantic subject to boys of that period and a symbol of new times and an unlimited future.

He had the normal traits and curiosity of boys everywhere. When he was 12, Walter snitched a handful of Virginia Cheroots from a showcase in his father's store, slipped behind a high fence with two friends, Ray Welch and Howard Pierce, and smoked them. It was the beginning of a lifetime cigar habit for Walter.

By the time he was big enough, he was playing sand-lot baseball for the Terrell school. There were pleasant hours to be filled, but Walter felt a sense of more important work to be done. He could see growth and change bubbling across the land. He was stirred by a restless spirit that meant he would soon get on with the task of shaping his life.

Chapter 2

There was not much in Walter Lechner's background to endow him with great fires of ambition. But from his Bavarian grandfather he had inherited the inclination to dream.

By the time he was 16, Walter had made the first of several major decisions. He loved the land, but it lacked fresh challenge. He was not going to be a farmer or a grocer like his father; he longed to seek new opportunities not available in a small town.

While in the tenth grade at Terrell High, Walter chose to bypass graduation and enroll directly at North Texas University, where courses in stenography and typing were available. He wanted clerical training to fall back on in the event he couldn't afford to finish engineering school, which was his goal at the time. It was a far-sighted and practical outlook for so young a boy.

The panic of 1907 shadowed his plans. His father lost his grocery store and went to work as a clerk in another store owned by R.W. Tickell. On Saturdays Walter also worked in the store making deliveries, waiting on customers and doing odd jobs. Walter doubted that he could attend college very long, if at all. He knew that with one brother and four sisters his family was going to need help.

At work in him now were the opposing forces of adventure and caution. He had decided to become an engineer while working one summer for the Terrell Electric Light Company. He had an instinct for fixing machinery and making tools and he used his hands with ease.

But while he harbored much higher ambitions for the future, he was confident that without the right kind of application, success could not be achieved easily. His idea of a successful man was Charlie Carter, a court reporter who earned $300 a month, probably more than any other working man in his community.

Walter learned his first shorthand from Carter. He taught himself to type on old Remington, Oliver and Blickensderfer typewriters. The Blickensderfer was a rotating head model, eventually brought back by IBM.

In college he studied mathematics, English, science and stenography. His precise mind, his attention to detail and a strong sense of thoroughness—plus his desire to be another Charlie Carter—quickly established him as an expert in stenography.

After his first year at North Texas, then a Methodist junior college (later known as Texas Military College), the Reverend Joseph J. Morgan, president of the school, approached Walter about teaching a class in stenography.

The course required a knowledge of English grammar, correspondence, spelling, civil government, commercial law, shorthand and typing. No one could graduate without being able to compose English with ease and accuracy.

Walter Lechner, not yet 18 years old, was being asked to teach a complex college course. He accepted on one condition—that he could satisfy both himself and the Reverend Morgan of his competence to do the job by taking an examination. He spent the summer preparing and passed his tests with ease. The Reverend Morgan returned his paper with a smile and said, "You'll do."

A month before he was 18, Walter Lechner was listed on the North Texas University faculty as head of the stenography department.

His year on the teaching staff was a fulfilling one, even though it provided little in the way of remuneration. North Texas was a small but elite school with an excellent reputation. It had a military as well as a religious base. Besides the Reverend Morgan, the men Walter knew best were Major Robert H. Clagett, the commandant, and L.H. Kidd, manager of the boys' hall. The school provided such courses as Bible, Greek and psychology, all of which the Reverend Morgan taught, plus courses in mathematics, science, English, music, voice, violin, art, Latin and modern languages.

An insight into the character of the times is contained in the honor code of North Texas:

. . . While we seek to be helpful in building character and forming correct habits, we are not a reformatory, and the student whose influence is unwholesome will not be retained.

The following rules will be enforced:

Students must be prompt in attendance, faithful to duty, and respectful to teachers.

All students, whether in the dormitory or in town, are required to attend chapel.

During study hours of the day all students must be in the study rooms, when not reciting.

Students will not be permitted to leave school, or absent themselves from school duties without the consent of the faculty.

No money will be loaned or advanced to pupils.

Smoking is not allowed, especially in or about the school premises, or in public. Those who are slaves to this habit are not wanted. Experience shows that their influence is hurtful, and unless such are willing to make an honest effort to stop smoking, we do not care for their attendance.

This last paragraph required Walter to conceal the cigar habit he had acquired somewhat earlier. It was a challenge he met with admirable success.

Most of the students in Walter's classes were older than he. But as the Reverend Morgan had predicted, he did well. All his students graduated and Lechner helped them find jobs.

By now he had completed his own post-graduate work and the time had come for new directions. He decided to enter the Texas Agricultural and Mechanical College to study textile engineering, intending to enlarge his knowledge of math, chemistry and physics. It was necessary for Walter to pay for most of his schooling. His father's salary and income from the farm were not sufficient to cover the lad's full expenses and support a large family.

So Walter found a job selling cars for a dealership in Dallas. Automotives was, in 1910, a relatively new field and a young man with energy and poise could create his own sales techniques. The line of automobiles he sold included the Thomas Flyer, the Apperson Jack Rabbit and the Detroit electric cars. Most were sold "barefooted." That is, their only extra equipment was the four wheels under them. The buyer had to pay extra for windshields, running boards, bumpers, headlights, horns, tops and practically everything else.

So quickly did he learn the business and so effective was his salesmanship that, before he left the agency, Lechner was promoted to secretary-treasurer. It was difficult for a young man not yet 21 to give up such a promising career in the automotive field, but he was determined to continue his college education.

In 1912 he entered Texas A&M, a school whose discipline and spirit made it unique among the campuses of the Southwest. It was a place where men suffered together in a womanless world, where friendships were formed that would last a lifetime. Lechner ranked high in all his classes. But after a year and a half at A&M, he was offered a job too good to turn down. Not only that, it almost fit the pattern of his earliest dream of a career.

He was offered a position with Texas and Gulf Railroad in Longview. Colonel L.P. Featherstone, the man who had built the Gulf and Interstate Line from Beaumont to Port Bolivar, had proposed a new Santa Fe line from Longview to Ore City in Upshur County. This line would be called the Port Bolivar and Iron Ore Railroad and would haul iron ore and coal from the mines of Northeast Texas to the sea at Port Bolivar. It was to be 30 miles long, reaching a point three miles south of the mine site in a small town called Ero, where it would join Santa Fe's Texas and Gulf branch at Longview.

For two years Walter spent some of the most exciting days of his young life working out of Longview. Then he was transferred to Beaumont to become private secretary to J.A. Glen, superintendent of the Gulf, Colorado and Santa Fe system. This was a more challenging job that enabled him for the first time to observe and study management methods. He became a close friend of Glen and other Santa Fe officials in both Beaumont and Galveston, the division headquarters.

His experiences with Glen, as his private secretary, were wide and varied. Once he built an inspection car for Glen to use on the tracks, an innovation later copied by many others in the business. He built it out of junk, including an old automobile engine that powered it. He frequently traveled the system in Glen's private car and learned the names and jobs of most of the men in the company. They were people he would never forget. Throughout his life, Walter Lechner would rarely forget anyone he met and few would forget him.

In the summer of 1915 Lechner learned of a job opening in Port Arthur in the export department of the Texaco refinery on Sabine Lake, north of the Gulf of Mexico and a few miles south of the historic Spindletop oil field. He was curious about the oil industry and attracted by it. So he jumped at the chance to get involved. For two years he handled the export billings until he concluded that Port Arthur was not big enough for both Walter Lechner and the mosquitoes that lived there. The insects were known locally as "galley nippers" and, to avoid their attack on his way to the refinery, Walter wore a Towers yellow slicker, laced boots and a mosquito bar over his head. In a slight Texas exaggeration, Walter told his friends that the Port Arthur mosquitoes were so big that "upland hunters would come down and shoot them for doves." Early in 1917 he decided to give the refinery back to the mosquitoes.

Not yet ready to give up on the oil industry, he decided to try his hand at the classic drudgery of the oil fields. He met one of the partners of the Cezeaux and Martin Drilling Company, which was then operating on Moonshine Hill in the Humble field in Harris County.

This was his introduction to the real world of oil. He went to work on the lowest job on a drilling rig, as a roustabout, one notch below a roughneck, but within a few months he had, as usual, applied himself, stayed sober and out of trouble so he was given a chance as a driller. This experience led him to several of the wildest boom fields in the great salt dome country of the Gulf Coast. For the next year he worked in salt dome oil fields at Batson's Prairie, Saratoga, and Sour Lake (all in the Big Thicket) and Goose Creek on Galveston Bay. That was enough oil field education for any young man still in his twenties. He learned to take a lease, put together an oil exploration deal, handle general oil field office work and to work with and handle field workers, landowners and oil operators, both independents and those in the major companies.

Great and unforeseen events were stirring thousands of miles away, events that would rearrange the future for Walter Lechner and thousands of young Americans like him. At a town called

Sarajevo, in Serbia, Archduke Francis Ferdinand, heir to the Austrian and Hungarian thrones, was murdered on Sunday morning, June 28, 1914. The two bullets that killed him signaled the start of World War I.

With the war in Europe closing in on the United States, Walter was ready to get into the fray as soon as his government declared war. To prepare for that moment, which seemed relatively near, he moved to Dallas and took a job with the U.S. Bond and Mortgage Company.

President Woodrow Wilson declared war on Germany and Walter put his business in order and prepared his parents for his entry into the service.

Chapter 3

Never before, and never since, has a nation sent its young men off to war with such a festive air and such a sense of purpose. In America they marched aboard troopships singing George M. Cohan's "Over There." It was "the war to make the world safe for democracy" and "the war to end all wars." It was also the war the President had vowed, in his election campaign, that America would never enter. Overnight it moved from the European War to the World War.

Walter Lechner heard the music and the rustle of the flag. He waited for his draft call, but when it did not come, he enlisted on December 11, 1917, in the Aviation Section of the Signal Corps of the United States Army. His choice favored the new and the bold. This would be the first war ever fought in the skies. It would usher in a new age of military tactics.

Lechner volunteered at Love Field in Dallas, under conditions no more attractive than those he would later encounter in Europe. The field was still under construction, with no company streets or sidewalks. The winter brought heavy rain and sloppy snow. Mud and slush were everywhere. The barracks buildings were incomplete and a labor dispute left the windows and doors without glass. The troops had no straw for their bedsacks and the Army issue of three thin blankets was not equal to human needs for warmth or comfort.

Still, the young men at Love Field were eager to join the adventure that rumbled on across the ocean. They followed the daily news reports and read the headlines and heard the rumors with a mixture of impatience and envy. Colossal events were unfolding all over the world. The first American troops, called "doughboys," had landed in France on June 26, 1917. In November, the Bolsheviks had seized power in Russia. In December, Jerusalem fell to the Allies.

At Love Field, the War Department separated the Aviation Section from the Signal Corps and formed the Aviation Section of the United States Army. The official insignia became wings superimposed over a four-bladed propeller.

Then a program of flight training began, using mostly civilian instructors. The Air Corps was so new that the field was staffed by former officers from the Cavalry, Infantry and Field Artillery, many still wearing boots and spurs. Instructors would utter with what they considered humor: "I haven't flown one of these planes, but I'll tell you how to fly them." Actually, the poor weather and the lack of flight ramps made little flight training possible. Planes bogged down in the muddy fields.

But events moved too swiftly to allow for frustration. One day the volunteers—in the Army idiom of the day they were known as "casuals"—were told that two squadrons were to be formed. One was to be known as Headquarters Company and would be based at Love Field. The other squadron was to be Number 169 and would embark shortly for overseas duty. The response was overwhelmingly in favor of service with the 169th, composed of men from many states.

Most of the men were motivated simply by patriotism, a fervent desire to serve the nation. It was an emotion that in another time, in what amounted to another world, would not be quite so much in fashion. And some of the motivation, of course, was a more basic urge to get away from Love Field, its mud and its strict regulations. The troops were in sight of the downtown buildings of Dallas, but were refused passes to visit even their homes on weekends or at any other time.

Six more men than were needed volunteered to form the 169th Aero Squadron and accepted assignment overseas. Walter W. Lechner was among them.

A few days later ranks and duties were announced. Lechner was installed as supply sergeant, in charge of clothing, ordnance and other equipment. He would wear the stripes of a sergeant first class until he was made a temporary warrant officer during combat duty.

The 169th moved by troop train to Roosevelt Field on Long Island, New York, where a scene greeted them that seemed to be out of the Russian front. Snow was falling and the young Texans would not see the good green grass of the United States again until their return from France.

At Roosevelt, one of those happenings occurred that leave men to ponder the larger questions of life. The squadron members were given their overseas immunization shots and the squadron baggage was marked for shipment to England. But during the final medical inspection, it was learned that a soldier named T.F. (Slim) Fortenberry had the measles. The entire 169th Squadron was quarantined until he was cured.

As fate would have it, the troopship that was to have transported the 169th to England was torpedoed off the coast of Ireland with the loss of most of those on board. A case of measles had, ironically, saved the squadron from disaster.

At last they arrived "over there," landing at Liverpool, England. They had been preceded by the American Expeditionary Forces under General John J. Pershing, whose arrival in France was announced with the dramatic words, "Lafayette, we are here!" as he laid a wreath on Lafayette's tomb.

But the American forces in England were not yet fully organized and the 169th Aero Squadron was attached to the British Royal Flying Corps, training pilots and ferrying planes across the English Channel to the front in France. From Liverpool, the squadron moved to Romsey, then to Flower Down, a part of a field 30 miles square known as Salisbury Plain, where Allied troops from all over the world were being trained.

From Salisbury the squadron was reassigned to Andover Field at Hampshire, England. Nine months later it crossed the English Channel to France, attaching itself for the next six weeks to the French Aviation Militaire.

The 169th then became a part of the Aviation Section of General Pershing's First Army, taking part in the "Big Push" that was to end the war on November 11, 1918. The squadron saw its first action at Saint Mihiel on September 12, 1918, the first distinctly American operation of the war. In the skies over France, General Billy Mitchell directed a massive aerial assault, the largest of the war, with 1,481 Allied airplanes participating.

As part of the Allied juggernaut, the 169th swept toward the Meuse-Argonne region. Over a million American soldiers fought during that advance, helping to break through the heavily fortified Hindenburg Line. One out of every ten was killed or wounded, a loss rarely equalled even in later years of more sophisticated weaponry.

Nearly five long years of conflict had introduced a new military vocabulary to the world: airplanes, zeppelins, machine guns, tanks, trenches, submarines, gas masks, liberty bonds. And in the front windows of homes all over America, flags appeared bearing blue stars for husbands and sons who had gone to war and gold stars for those who would never return.

On November 9, 1918, Kaiser Wilhelm abdicated and fled to The Netherlands. Two days later, in a drizzling rain in the Compiegne Forest, the armistice was signed, ending the First World War. Marshall Ferdinand Foch of France, Supreme Commander of all Allied forces, signed for the Allies.

Walter Lechner and the 169th Aero Squadron left Europe and arrived back in the United States on May 1, 1919. They had been gone for 15 critical months that changed the world.

It was good to be home. In their haste to get there, the men of the 169th had not allowed themselves to be detained by unnecessary cargo. One of Lechner's vivid recollections of their departure involved a quartermaster store at Romaratin where no one seemed to know what to do with hundreds of new, unpacked typewriters. Finally, they were removed from the crates, tossed into a pile and destroyed. Walter watched that great waste in total disbelief, as a man whose education was steeped in writing machines. The boxes were used for firewood.

Most other troops had preceded the 169th in returning to America. They were frequently moved about in France for several weeks after the armistice. Finally, they moved out on the S.S. Henry R. Mallory, sailing from Bordeaux and landing in New York Harbor near the berth they had left from. Walter Lechner's duty was to check out all the Signal Corps troops coming back from Europe He spent several weeks at the depot located in Mineola, New York.

Lechner was discharged at Mitchell Field on Long Island, New York, on May 16. He was awarded a Bronze Star medal containing four bars, each one citing meritorious service during the battles of the Meuse River, the Saint Mihiel Salient, the Argonne Forest and the Defensive Sector.

As the reports of the war kept coming in, he learned more of what had gone on in France than he had ever suspected while he was in the middle of it.

He was surprised to learn that on September 12 and 13 American forces attacking on both sides of the St. Mihiel salient had squeezed the German troops, forcing more than 15,000 of them to surrender on those two days. That was the beginning of the end for the invaders.

Six major prolonged assaults were made on the Germans. Two of these at St. Mihiel and the Meuse-Argonne Forest in which 1.2 million men had been engaged. During the war 115,000 Americans were killed, 206,000 wounded and 4,500 captured. Armies of all nations on both sides lost more than 10 million and twice that many wounded. Germany and Russia each lost almost 2 million and France, Austria-Hungary and Great Britain each lost more than a million men.

The war's end was signaled by a mutiny in the German fleet at Kiel on October 28. It spread rapidly to Hamburg, Lubeck and Bremen and then to the whole of Northwest Germany and the naval war was over.

On October 27 General Ludendorf's resignation had been accepted and he was succeeded by General Von Groner. On November 7 and 8 the revolution broke out in Munich. General Von Groner and Marshall Hindenburg informed the Kaiser they could no longer guarantee the loyalty of the army. The next day the Kaiser abdicated and on November 10, heeding the advice of his generals, he fled to Holland. The next day, the Armistice, on which negotiations had begun while the mutiny was going on in Munich on the eighth, was signed.

Those, Walter learned, were the major events leading up to the end of the war. Now it was all over. It was the war to end war and the war to make the world safe for democracy. That's what President Woodrow Wilson had promised. Now the world would have to see if it was worth all the carnage, slaughter, sacrifice and suffering of four long years.

So Walter Lechner was home again. He was 29 and one half years old. He was a muscular, vigorous, wiry young man, 5 feet, 9 inches tall with a ruddy complexion, brown hair and brown eyes, and with something picked up in France—a moustache that would remain. A cigar was constantly in his mouth.

For Walter the war was a time of great experience and excitement as well as an opportunity for service to his country. But it was also the beginning of friendships that would continue through the remainder of his useful and challenging life.

While Walter was boning up on the war he had just finished fighting, he was also being inspired by the glowing reports from the oil fields of North and North Central Texas. Booms such as Ranger, Desdemona, Breckenridge and Burkburnett were all in full bloom. Since Spindletop, the world had never known such booms. The eastern newspaper accounts made them even more colorful than they actually were. But Walter was fascinated. Now he knew this was the business he wanted to get into and stay with permanently. He promised himself if he could get into the oil business, he would never want to get into anything else. From what he was reading about the big booms, that was all the opportunity and excitement any man could ever want.

One day while he was thinking about how he was going to get a new start in the oil business, he received a message from an oil company in Burkburnett. The company was Currin and Kean. Walter had heard of the partners. They had heard of him too, and since good oilmen were hard to find, they sent for him as soon as they learned his address.

With only a few days left to finish his assignment at Garden City, Walter sent a telegram accepting the job and stating that he would travel directly to Wichita Falls and report to their office there as soon as he checked in his last soldier. He did that on May 16, 1919, and he was severed from the service. He didn't even go home to Terrell to see his family before reporting to Currin and Kean's Carnation Oil Company for work on May 19.

One thing that impressed Walter Lechner about the oil industry was the fact that a man could start from absolutely taw and work himself to the top in a short time if he had the ambition, the guts and the luck. Going into oil didn't require any investment or formal education. It did require a willingness to work long hours without compensation and to live on one's wits for a long time. And it required a native intelligence composed largely of common sense. These were assets Walter possessed.

Chapter 4

It was the thirteenth of May when Walter received the telegram from Currin and Kean offering him an immediate job in the expanding Burkburnett field 14 miles north of Wichita Falls. Thirteen had been his lucky number since the day he was born. Walter had almost finished checking out his old buddies, so he went to his commanding officer, Colonel W.L. Moose, and asked what he should do about replying. The colonel told him to say he would be there in a week. On May 16 Walter had his honorable discharge and that evening was on his way to Texas. He rode the Baltimore & Ohio Railroad to St. Louis and the Missouri, Kansas & Texas from there to Dallas and on to Wichita Falls.

He was astonished at his first sight of Wichita Falls, a booming center of activities south of Burkburnett which was probably the wildest oil boom in the history of great Texas booms. The last time he had seen the town, it was a sleepy little supply center and water hole for farmers and cattlemen. Now it was bristling with business. There were good hotels, paved streets, hundreds of automobiles, cafes and most of the other activities of the oil country. Since August, its population had grown from 17,000 to 25,000.

Currin and Kean were delighted to see Walter. They lost little time in telling him about their operation and the boom in Burkburnett and then hustled him out of town in a new car they had ready for him.

When he drove into Burkburnett, he was mesmerized by the color and excitement. Day and night, without letup, the streets were filled with automobiles, wagons, horses, buggies, mules, oxen, any form of transportation. The Katy Railroad seemed to have trains coming into town hourly and unloading men and materials for the field. In the last month three new special trains offered daily service. In 1919, the Missouri, Kansas & Texas (Katy) line was handling 3,000 passengers daily. Late afternoon Pullman cars to Fort Worth were filled with men who had return tickets for early morning.

Everywhere he looked, there were derricks towering over rigs drilling for oil, and even more already-producing wells. There were hundreds of stores, most of them new. There was an oil well on almost every lot in town. All drilling was done with standard rigs. There was none of the rotary drilling Walter had learned in Humble.

Burkburnett was a poor town before the oil boom. One drought after another had hit the farmers in this town near the Red River which marked the dividing line between Texas and Oklahoma.

The railroad was building a new depot, freight facilities and loading terminals.

At night people were sleeping in chairs in cafes and hotel lobbies, on front porches of homes and in pews of churches.

Rain would convert the town into a quagmire with vehicles and animals bogged down for miles in all directions. A bus line that had started operating a month earlier to relieve the passenger traffic jam was of little use during rainy weather.

It was amazing to Walter to see a church with a slush pit at its back door and another with a well pumping oil in its front yard. At night electric lights on rigs twinkled all over town and deep into the field. Work never slowed in the field and gambling dens ran full tilt 24 hours a day, some right alongside schools and churches. Prostitutes ran around town in kimonos and flimsy nightgowns. Promoters and con men and their gullible victims swelled the mob.

Holdups and murder were commonplace. Despite the new prohibition law, there were a hundred watering holes in town. They were called "blind tigers," usually in back rooms of cafes, pool halls or even ice cream parlors.

Returning soldiers and sailors, most of them still wearing all or part of their uniforms, were everywhere. All of them were either engaged in oil field work or had good jobs in some of the stores and shops or other businesses in town. Some had come home to drill wells in their own yards or on their farms.

The lumber for housing and derricks came into Electra by train and was sent on to Burkburnett by wagon and team. Trucks and automobiles were almost as helpless as buses on the deeply rutted streets and roads. Only animal-drawn vehicles and horsemen provided reliable transportation.

On the school grounds was one well and three or four more were being drilled. All the teachers were dabbling in oil stocks and the pupils were learning more about producing oil than about reading, writing and arithmetic.

The publishers of the *Burkburnett Star*, the Laney brothers, were having a difficult time meeting the demand for advertising because they simply could not get enough printers to get the paper out. Stock promoters, many of them women, were hawking their shares in the streets and hotel lobbies, and cleaning up. Labor was in short supply since there were few places men could sleep. Most boarding houses were selling beds and meals in three shifts a day. The same was true of the hotels. Pool halls sold tables as beds with as many as two and sometimes three men sleeping on a table after midnight. Even the honky-tonks that closed at midnight sold chairs for sleeping after hours. After working hours, prostitutes sold half their beds to weary workers interested in nothing but sleep.

Walter was fascinated and amazed at the sights he saw. Somehow Burkburnett brought back memories of the Meuse-Argonne and other battles.

What attracted many of the mob were stories of a few lucky men such as Schields Fowler who struck it rich with oil. In 1918 Fowler was a disgusted farmer about 50 years old. Pure fate led him to one of the great Texas fortunes. He spent his boyhood in Lincoln County, Tennessee, where he was born on November 9, 1867. He and a friend went to Alvarado, Texas, in their teens, arriving penniless after losing their money in a crooked card game in Greenville where they stopped to change trains.

Fowler had to take any job he could get, so he went to work for a farmer named Shropshire and married his daughter Cassie. After farming in various areas of Texas, Fowler and Cassie moved to Burkburnett with their two sons and four daughters shortly after the turn of the century. A born trader, he bought one farm, didn't like it much and traded it for another with J.N. Vaughan. On this place he finally struck it rich.

Schields Fowler wanted to trade that farm, but Cassie demanded that he drill one well on the land before she would sign the papers. Dry holes east, west and south of their farm did not dampen Cassie's belief that there was oil beneath their land. What convinced her was the oily taste of the water from the family well.

Fowler was so determined to get out of Burkburnett that he raised $12,000 to drill a well after Magnolia Petroleum Company turned down his offer to take the venture on a 50-50 basis with him.

He sold investment shares in his Fowler Farm Oil Company for $50 to $100 each, but didn't raise as much money as he needed. He got his own cash by selling a load of cattle. Then he got Walter D. Cline of Wichita Falls to come in with him and bring his drilling rig. Between them Fowler and Cline had the lion's share of the stock. They hired Lynn Granberry to drill the well and set their target depth at around 1,700 or 1,800 feet.

In the early morning of July 26, 1918, Red McDowell, a member of the drilling crew, came running to the Fowler farm shouting that oil was flowing down the cotton rows and ruining everything. After bringing the well under control, Fowler measured the flow at some 3,000 barrels of oil a day. Soon it was being estimated that the well would go to as much as 10,000 barrels a day. It was a major oil discovery.

Within a few days, drought-stricken farm lands and run-down town lots that could have been bought on July 25 for a few dollars were being leased to oilmen for fabulous prices. Poor landowners were rich days after the well they called "Fowler's Folly" came in. Prosperity spread across the county like the German army going through the low countries.

Ironically, the Magnolia Petroleum Company, the Standard Oil subsidiary which had refused to help Fowler drill his well "until he showed them there was oil there," paid him $1,800,000 for his property, plus royalty.

That was the kind of story that caused Walter Lechner's eyes to brighten as he listened. If Schields Fowler, a 50-year-old dirt farmer, could make a fortune in oil, anyone could do it—

especially if the big oil companies continued to exhibit such lack of vision, imagination or willingness to venture. It seemed the big companies avoided the gambles and just waited to buy out the little operators who took all the chances. Of course, only one little man in a thousand had the luck of Schields Fowler. The big boys played close to their vests, moving out their stack of chips only when they believed they had a cinch.

Even the name of Burkburnett had color. The town was named after pioneer rancher Burk Burnett who owned the vast 6666 (Four Sixes) ranch. And it was the locale for Rex Beach's famous oil novel, "Flowing Gold," based on "Fowler's Folly."

When Walter reached Burkburnett, there were almost 200 wells flowing and another 200 being drilled. There were 300 stock companies on the Wichita Falls oil exchange, where Walter also did some trading in the Carnation Oil Company and the Dixieland Oil Company. By May the investors who had put money in Fowler's deal had already collected $1,750 for each $100 they put up. During 1919, the field's best year, Burkburnett produced 31.6 million barrels of oil, which was a great contribution toward relieving the nation's oil shortage which had already seen gasoline selling for as high as $1 a gallon in some areas although the average price was far below that.

As wild as Burkburnett was, it still had company in the Electra field less than 20 miles to the west.

And the really gigantic oil play in Texas, or the nation for that matter, at the moment was about 75 to 100 miles south where three major booms, all rolled into one, were astounding the world of petroleum. As if on some schedule, they had been discovered starting late in 1916, even before the United States got into the war. First was Breckenridge in Stephens County. Then in 1917 the great Ranger field in Eastland County came in. Eleven months after that, the Desdemona field came in near Hogtown at the juncture of three new counties—Eastland, Erath and Comanche.

All of this activity, of course, was in full bloom, along with Burkburnett, which had come in after Ranger and before Desdemona.

Breckenridge, 35 miles north of Ranger, came in on October 30, 1916. The discoverer was Producers Oil Company, the producing arm of Texaco (then known as the Texas Company). Producers had bought up some 50,000 acres of 10-year leases at 10 cents an acre in 1911. The well was on the B.J.W. Parks ranch. But this and several other wells drilled later were not much to shout about. In fact in 1916 the production was negligible. In 1917 it was only 36,000 barrels and in 1918 production was almost a million barrels, but that was still somewhat less than a giant oil field. It was not until 1919 that Breckenridge came into its own, flowing some 10 million barrels of oil. In 1920 that figure more than doubled and by 1921, its peak year, the flow went up to more than 31 million barrels. After that it tapered off and paid the penalty of close, town lot drilling which dissipated bottom hole pressure and brought on early salt water intrusion, leaving hundreds of millions of barrels of oil behind never to be produced. As an example of other waste, the field started producing gas, for which there was no market. The practice then was to let the wells flow wide open until they blew off the gas and the oil started to flow.

Breckenridge was largely a big company field. Besides Texaco, the big companies included Humble (now Exxon USA), Gulf and others.

In the summer of 1916 Breckenridge was a town of 600 people. After the Parks ranch well came in, population stepped up to 1,000, and by the big boom days of 1919 and 1920 Breckenridge was a relative metropolis of 40,000. Cowboys stopped cowboying, farmers left the farms, and school teachers and clerks became field hands, prospectors, promoters and lease brokers. It was a beautiful field while it lasted, producing from four different formations.

Ranger was the next bonanza to get started. That was on October 25, 1917, about the time Walter and his Signal Corps unit were moving out of Love Field in Dallas. Ranger was a relatively sleepy little cow town on the road from Fort Worth to Abilene until it was roused out of its slumber by a tremendous gusher which blew in.

The field was found by a talented and determined young surveyor named W.K. Gordon. He had arrived in Texas in 1913 to survey a railroad right of way. He found coal and formed a coal company which became Texas and Pacific Coal and Oil Company in time. This pleased his old Texas and Pacific Railroad since it brought them tonnage. The blooming of the Electra field the year before World War I inspired him to seek oil near Ranger. His first well was a mechanical failure, but he was in a position to raise money for a second well. The test was on the J.J. McClesky farm in the center of a 25,000-acre lease block the citizens had rounded up for him as compensation for drilling the well. In late October Gordon's New York office gave him instructions to cease drilling. He ignored the orders and went ahead about 200 feet deeper where he hit the rich Marble Falls sand. The boom was on.

Edward L. Doheny, a colorful Wisconsin promoter later indicted for bribing Secretary of Interior Albert B. Fall in the Elk Hills and Teapot Dome oil scandals, was active in Ranger in 1919. From there he made a prophetic statement for *Hearst's Magazine* which Walter read while he was still in Mineola, New York. He said:

"Ere long oil-driven machinery will enable man to conquer the air, till the soil, utilize the seas, accelerate communication on a scale not dreamed of today. We are about to enter the oil age— the age of motorization."

This pronouncement by a man who had discovered the first oil in Mexico, found oil in Venezuela, formed the great Pan American Petroleum Company and blazed his way across the nation with one important oil discovery after another was fuel for Walter Lechner's determination to become an oilman as soon as he could get out of the army.

About the same time the president of the Standard Oil Company of Indiana predicted, "The great concern for the future would be high grade petroleum to meet requirements in the manufacture of fuel for gasoline engines."

This opened Walter's eyes even wider and underscored his belief that oil was the industry of the future. He reasoned that the nation's future was unlimited if sufficient oil to provide energy for its potential growth could be found. He told Colonel Moose it was his intention to do his part in this fight for the future just as he had done in war in France, and the colonel believed him.

Ranger and Breckenridge were only two-thirds of the three-headed oil region of North Central Texas. The third was

Desdemona, once called Desdemonia until the post office department in Washington changed it. Even before that, it was known
as Hogtown, a name a few of the inhabitants never changed. It
was called Hogtown because it was on Hog Creek, one of 14
streams in Texas of the same name. Hog Creek was lined with
post oak trees and hogs were permitted to eat acorns along its
banks. Its banks were also slick with oil seepages.

For about 20 years, signs of oil had been noted throughout
Central Texas. There were many seeps as well as outcrops, oil-
saturated soil and tainted water, but little geology was practiced
in the area. This changed at Desdemona. R.O. Harvey of
Wichita Falls sent W.E. Wrather, later to become one of the
most famous petroleum geologists in the world, and a lease expert named L.H. Cullum to the Hog Creek area to make a
geological investigation. Wrather found an anticline as he
studied rocks along the creek bank and reported his findings to
Harvey. Immediately Cullum met with a group of local citizens
and agreed to lease 6,000 acres of land. The citizens group called
itself the Hog Creek Oil Company.

In the meantime the prospect that led Harvey to Hog Creek
was abandoned. It was a well being drilled by Tom Dees of
Oklahoma on a 200-acre block. The Maples Oil and Gas Company agreed to drill a well on the Joe Duke farm for half interest
in a 5,000-acre block of leases. Wrather, incidentally, picked that
location also. This was the well that finally blew in as a gasser
and turned into a 2,000-barrel oil producer at 2,960 feet. The
flow was from a black lime formation. The well got better with
time, but an even better well, one flowing 8,000 barrels of oil daily, came in shortly after that. The third well was drilled by a syn-
dicate which included the world's greatest wildcatters of the time,
Benedum and Trees of Pennsylvania.

That well blew the lid off things in North Central Texas. People from Breckenridge, especially those that had not been in on
the big stuff in the beginning, started flowing toward
Desdemona.

Soon the lure of fabulous wealth reached the outside world and
camp followers by the thousands started rolling into Desdemona.
When the well came in, Desdemona was a poor, stagnant village
about ready to become a ghost town. A blacksmith shop, a cot-

ton gin and a post office formed the main business district. But its people were high in their faith in oil. Oil in varying quantities had been found from Petrolia to Strawn and Desdemona was on a direct line between the two. Hog Creek was in good watermelon, fruit, cattle and horse country. The families were large and the people were healthy, energetic and God-fearing.

When oil came in, Desdemona was a typical oil boom town. Building was accelerated to take care of the incoming mob. It was a little man's boom in contrast with Breckenridge where the big companies were in the saddle. Bankers, merchants, manufacturers and other businessmen, ignorant of the hazards and risks of oil, came in and were unable to resist the spiels of the promoters. Most of them lost all they came with. A few got richer.

The biggest incident of the Desdemona boom was the famous gas well on the Payne farm. It blew in flowing from 30 to 40 million cubic feet of gas into the air daily for months before it blew itself out. The noise could be heard 25 to 50 miles away and the waste (in modern terms) was disgraceful. But there was no market for gas then. Also disgraceful was the waste throughout the entire Desdemona-Ranger-Breckenridge triangle. Wells were permitted to flow over their storage tanks, even turned on to gush for victims of the bogus stock promoters. The natural waste was due to close spacing and uncontrolled flow. The result was that more than 90 percent of the potentially recoverable oil was never to be recovered and all the natural gas was blown into the air.

North Central oil, that is Ranger, Desdemona and Breckenridge (as contrasted with North Texas oil in and around Burkburnett), produced 20 million barrels of oil in 1918 but more than doubled that in 1920 up to 53 million barrels. Due to the burgeoning demand caused by the drain of the war and the post-war expansion, the 1919-1920 prices for oil were the best they would be for another 37 years. It was this demand and the resulting high prices which turned the industry around. In the next few years, more oil would be discovered than the country could use, a condition which continued through the early 60's. These were the years of cheap oil, high exports and an almost total disregard for the true value of natural gas. The country was using all the oil it could produce, but gradually the incentive for the wildcatter diminished and the result was the fading out of the old wildcatter.

Walter visited all these fields, invested in most of them or worked in them with Currin and Kean or one of their companies. He came home from war and walked right into a baptism of fire in wildcatting in one of the greatest booms in the country's history.

Currin and Kean told Walter his job would be simple—just do whatever it took to run the company's field operations, even when that meant working on the derrick floor or making deals.

Some of his time was spent reading the *Oil Weekly* out of Houston. He read geological articles by such notable geologists as Dr. F.H. Lahee, former professor of geology at Massachusetts Institute of Technology. One particularly interesting article dealt with how geologists look for petroleum, including fundamental facts about oil exploration that would serve him well in his chosen career. Closer to home was another article by Wallace E. Pratt, chief geologist for the modern and expanding Humble Oil and Refining Company, that dealt specifically with North Texas geologic structure and oil areas.

The stock promoters in the field were either good or bad. There were many swindles going on and the oil ignorant were falling for them. They ran full page advertisements in the daily newspapers, but the oil trade journals, such as the *Oil Weekly* and the *Oil and Gas Journal,* avoided them.

One they might have carried was an ad promoting a company being sponsored by "Farmer Jim" Ferguson, recently impeached governor of Texas. The headline said: "If you do not want to pinch a wildcat's tail, do not read the rest of this advertisement."

The body of the ad said:

> If you can afford to lose $100 and want to take a chance in a legitimate speculation, write to me. If you can't afford to lose the money, do not invest, because we do not know whether there is oil on my land or not.
>
> Most of these big oil advertisements are fakes and swindles, and when a fellow tells you he is dead sure he is going to hit oil, even in Ranger or Burkburnett, he is just dead sure knowingly telling you a black lie to get your money, and most generally he has so damn much water in his stock there will never be any room

for oil in his well. If you struck oil your interest would be one fifteen thousandths part in a piece of land that ain't big enough to cuss a tomcat.

As an honest man, I can only promise now that I will give my personal attention to the drilling of the well and that your money will not be paid out to stock salesmen, but will be used in drilling a hole in the ground.

Jim was even honest in the name of his company. It was Chance to Lose Oil Company.

Walter thought the old coot, husband of the future first woman governor to be elected, had a good sense of humor and thought he would get some response to his ad, which he certainly did.

There was another ad in the *Oil Weekly* that caught Walter's eye. It was placed there by the Dallas Chamber of Commerce and the Manufacturers' Association. After bragging about Dallas' mammoth population of 154,923, it claimed the city was the hub of the oil industry in the great Southwest. It had arrows pointing to Burkburnett, 160 miles; Ranger, 127 miles; Brownwood, 174 miles; Houston, 243 miles; Beaumont, 285 miles; Shreveport, 188 miles; Tulsa, 227 miles; and Healdton, Oklahoma, the most recent big discovery, 160 miles. These were the hot spots of the oil industry. But Fort Worth, Wichita Falls and Tulsa were getting the big oil play and Dallas was attempting to move in. And it was an effective ad. Walter decided that would be the place where he would eventually land permanently so he could operate all over the country with ease. He had no doubt that somewhere, some day, he was going to hit it big. Then he would make Dallas his permanent home. After all, it was only 31 miles from Terrell and his family.

These were days of some sadness for Walter also. Shortly after he arrived in New York from France, he was informed that his young wife had entered suit for divorce. He decided not to contest the action since it had long been obvious to him that he and his wife were almost totally incompatible, although she had given birth to a son. She would, according to the agreement, retain custody of the boy.

Now Walter was unhappy over the event—not that he didn't know divorce was right for both of them, but he had never in his life considered divorce. And worst of all, it would be the first divorce in the long history of the Lechner family.

In Burkburnett, though, he was sufficiently busy to keep his mind off such trouble. Currin and Kean had not only given him a job, but had also invited him to invest in any venture they entered, to do outside work for others when he had time and even to participate in ventures outside as well as with his company. This was the beginning of a very exciting and happy business experience.

At one time he took part of a deal with his bosses in Burkburnett and made a few thousand dollars. He also got a good bonus for selling a 125-acre tract in Desdemona to the Jenkins brothers, the owners of the Tennessee Oil and Gas Company.

But most of his days were spent around the field running Currin and Kean operations. The cable tool rig was something he had to learn to operate. In the Gulf Coast most of his work had been with the rotary. But in North Texas the rotary was not equal to the hard rock formations.

He learned all the legal points of leasing and securing titles from a young man just out of the Navy. His name was James V Allred. There was no period after the "V" because it wasn't an abbreviation. He had no middle name, so the Navy, whose regulations required a middle initial, gave him one.

Jimmie Allred, as he was known in his hometown of Wichita Falls, got his law degree in a one-year crash course at Cumberland University in Tennessee. The law school lawyers looked down on young men from short-course universities such as Cumberland. But probably most of them were happy Jimmie Allred had not gone the full course at the University of Texas. He beat them in court enough with his bob-tailed degree. And later he did it as a prosecutor and district attorney and as attorney general of Texas. Then he became governor of Texas for two terms, a federal judge and almost a United States senator. Most of his friends believe if he had beaten W. Lee O'Daniel, an erstwhile friend, in the Senate race (which he came very close to doing), he would have been President Roosevelt's choice for vice president, and maybe President of the United States.

Allred and Lechner were immediate and very close friends. They gave each other a lot of useful advice over lunch or dinner in the Kemp Hotel or one of the Wichita Falls or Burkburnett cafes, such as they were in those days.

Walter made many other friends in his six or seven months in the booming northern and north central oil fields. He also made about $30,000 above his salary on various deals, found a good well in Sipes Springs and sold it for a handsome profit.

Later he lost most of his money. When he drilled a dry hole that took his fortune down to about $6,000, he decided it was time to move on to Dallas and look for better battles to fight.

Chapter 5

Dallas was where Walter Lechner felt most at home. The city had 154,000 citizens and some said half were in the oil business, either directly or indirectly. Walter's immediate purpose in moving there from Wichita Falls, where he had been attached to the headquarters office of Currin and Kean, was to have an appendectomy. His company, whose fortunes were dropping somewhat as the great booms of North Central Texas began to die down, decided to open a Dallas office. That's where Walter reported back to work when he left the hospital.

The Currin and Kean office was in the Borger Building in downtown Dallas. Walter had a desk in an unpretentious office. Most of his work was outside, looking for deals or joint ventures for Currin and Kean and/or for himself.

He rented living quarters in the new and comfortable Dallas Athletic Club which catered largely to bachelors. It was a pleasant place to live, far removed from the boom towns and even the "luxury" of Wichita Falls. He met many new oil friends at the club since about half the club's residents were in the oil business one way or another.

A few months after Walter returned to Dallas, he met Ruth Nowlin from Glen Rose in Somervell County, south of Fort Worth. She was attending Miss McBride's Private School in Dallas after graduating from Glen Rose High School.

Through a secretary in the Currin and Kean office, Walter also met her sister, Marie, who was a secretary for the Atlantic Oil Company. Shortly after Ruth and Walter met, Ruth became a secretary for the Alamo Electric Company. Occasionally they would date and usually Marie and a friend of Walter's, Ray Hubbard, would double-date.

During the passing months, Walter revealed that he had been married before, but that didn't upset his friendship with Ruth.

Then, despite the fact that he lived quite well at the DAC and drove a nice car and always seemed to be able to pick up the movie and dinner checks, Walter admitted one night he wasn't a wealthy man. In fact, he said he was really on the lookout for a good job since Oliver Currin and Luther Kean were thinking about dissolving their partnership. Bad days had finally hit the company as their wells in Burkburnett were hurt by close spacing and unrestrained production.

This was an uncertain time for Walter. For the first time since he left high school (except the year he attended Texas A&M College), he faced enforced idleness. He had money left from his fling in the Burkburnett and Desdemona booms and was not averse to a rest. He was 30 years old in 1920 and felt it might be time for a breather. He still could make lease plays or even look around for a new base of operations. Regardless, he wouldn't hurry.

In the meantime his courtship of Ruth Nowlin was progressing quite well. One night he proposed but told her he was not yet financially secure.

"Don't let that worry you," she said. "You made it once and you'll make it again."

With that assurance of faith, Walter made arrangements for them to be married by Dr. Wallace Bassett, the pastor of a Baptist church in the Oak Cliff section of Dallas. The ceremony was performed in the parsonage at 1110 West 10th Street at 8 p.m. on September 1, 1920. Walter's friend, Guy Radley, was the best man and Ruth's youngest sister, Neville, was the maid of honor.

They skipped the honeymoon and moved into a furnished cottage on Harwood Street, which Walter had rented a week earlier. Ruth kept her job at Alamo Electric and Walter continued his activities as a lease broker. The oil business was in slow motion with the big fields running out of steam. Despite this, the Lechners were able to live fairly well. Eventually they found keeping the cottage too much with both of them working and moved into a room in Mrs. Pratt's home on Beacon Street.

In the spring of 1920, Currin and Kean decided to dissolve their partnership and terminate their firm.

This gave Walter time to think things over as he adjusted to total self-employment. There was never the slightest thought in his mind about turning to some other work. Oil was his business and in oil he would stay. He knew great opportunities were ahead. For instance, only a year before Walter's marriage, Harry Sinclair had formed a company that was beginning to move rapidly ahead. Another bright young man named W.G. Skelly had incorporated the Skelly Oil Company in Tulsa less than a month later. Both seemed to be getting off the ground in grand style.

He recalled a significant incident that had happened in Burkburnett in June, 1919, when the Texas Railroad Commission had organized its oil and gas division in compliance with a new state law and had gone to work immediately. On July 11 the commission had issued its first shutdown order in the Burkburnett field to attempt to stem the rape of the great reservoir there. A week later it had issued its first proration order in the northwest extension to Burkburnett. On July 26 it had issued its first rules relative to field development, including the famous Rule 37 to protect small tract landowners. This was a beginning in orderly oil and gas development and Walter was proud that he had been present when it happened. It was history in the making.

Almost at the same time the largest gas well in the world blew in as a gusher 25 miles north of Amarillo in Potter County, Texas, with a flow of 71 million cubic feet of gas daily. There was news of the Consolidated Gas Company in New York achieving a new send-out record of some 160,327,000 cubic feet of gas on a single day with the temperature at zero. These incidents, he reasoned, were signs of even greater opportunities in petroleum. Not only that, but in 1919 the oil industry had produced its five billionth barrel of crude. There was no stopping this industry.

In 1920 at his old Texaco refinery down in Port Arthur in the extreme southeast corner of Texas, the first Holmes-Manley cracking units were put into operation, opening a new era in refining. In July, 1920, the first Permian Basin oil was discovered when Underwriter Oil Company's No. 1 Abrams well came in as a small producer near Westbrook in Mitchell County. A notice in the *Colorado City Record* heralded the opening of what turned out to be one of the greatest oil and gas producing areas of the nation.

Walter and Ruth had been married less than three months when the colorful Col. A.E. Humphreys, with the backing of his old friend, "Black Jack" Pershing, Walter's World War hero and commander-in-chief, brought in the famous Rogers No. 1 well at Mexia. It was completed at a depth of 3,060 feet in the soon-to-be-famous Woodbine sand. The well also opened the great fault line play and a new era in Texas oil.

Walter, who had seen a half dozen potentially great oil fields ravaged by greed at Burkburnett, Desdemona, Ranger and other places in Central and North Central Texas, concurred when M.L. Requa, vice-president of Sinclair, cited the need for voluntary oil conservation. He said that, unless that was achieved, the states and the federal government would force such conservation on the industry.

"That's sure as hell right," Walter told Ruth after reading the item in the *Dallas News*.

An item which shocked Walter was a report quoting Secretary of the Navy Josephus Daniels as advocating nationalization of the oil industry to protect the nation's future on the seas.

For the first time in history, the term "petroleum engineer" was used when E.G. Wagy was appointed to such a position with the Standard Oil Company of California.

A week or so later, the City of Omaha announced that the gas company it had recently taken over was making gas from corn cobs. "That," Walter thought, "was another great use for cobs."

It was a fascinating era of American history. In 1920 the population of the nation had reached 150 million. Life expectancy had reached 54 years, up about 5 years from 1901. Both the Socialist Labor Party and the Socialist Party held conventions in New York to challenge the Republicans and the Democrats. In Chicago the Farm Labor Party held still another radical political convention. The Republicans became the party of the "smoke-filled room." The term is attributed to Harry M. Daugherty, a Republican from Ohio, who said, "The convention will be deadlocked, and after the other candidates have gone their limit, some twelve or fifteen men, worn out and bleary-eyed from lack of sleep, will sit down about two o'clock in the morning around a table in a smoke-filled room in some hotel and decide the nomination. When that time comes, Harding will be selected."

The statement was perfectly correct and the "smoke-filled room" has lived in political infamy. Walter's diary indicated he thought it was probably true, but a statement better left unsaid. A staunch Democrat, he considered the statement a typical Republican blunder.

That was the year the "Roaring Twenties" began. There were about 15 million cars, one to every four families in the nation. Man O'War won the Belmont Stakes and the Preakness; the U.S. won the Olympics in Belgium. In 1920 a grand jury indicted the Chicago White Sox for throwing the 1919 baseball World Series with Cincinnati, but a jury not only cleared the players, but carried eight of them from the court room on their shoulders. Cleveland defeated Brooklyn in five of seven games for the 1920 series. Franklin D. Roosevelt, running for the vice-presidency on the ticket with James M. Cox, was soundly defeated by Warren G. Harding.

Great writers on the scene included F. Scott Fitzgerald, Eugene O'Neill, Sinclair Lewis and Edith Wharton. John and Lionel Barrymore were in the cast of "The Jest" on Broadway. Illiteracy in the U.S. reached a new low of 6 percent. The year 1920 was a time of transition in American life.

One day in early 1922, Walter was stopped in the lobby of the DAC by E.I. "Tommy" Thompson, who had a proposition.

"What are you doing now?" Tommy asked Walter.

Walter thought a minute and found himself without an answer, so he told Tommy he was chasing a few leases and catching up on his reading. Tommy asked if he were interested in a job and an interest in a new company he was forming. Walter would put together a lease block and it shouldn't take more than three to six months, Tommy said.

"I am foot-loose and fancy-free, so I'm interested," Walter replied between puffs of his big black stogie.

Tommy explained that he and Buddy Fogelson had formed an exploration and production company and named it Loutex to indicate they would operate in Texas and Louisiana. He said they had some other associates including W.A. Reiter, Earl Sneed, W.H. Foster and Warren Oaks—all allied in the past with Colonel Humphreys of Denver. They wanted Walter to join their firm and maybe become their point man and superintendent and move to Snyder in Scurry County to take charge. They would give him $300 a month (a 1922 small fortune) and an interest in the company.

Walter thought a minute or so, then said, "Tommy, you've just made yourself a deal."

The only thing Walter worried about was Ruth. He wondered if she would quit her job and go out to West Texas with him on a wild goose chase. There was no oil in Scurry County and his job would be to find some. It wasn't any cinch, but he decided Ruth would go, so he accepted. He was weary of traipsing up and down the streets of Dallas and Fort Worth and to every small town, farm and ranch in between, gathering and trading leases and other interests. Things were getting slower every day, so this opportunity was just what he wanted. Tommy and Walter agreed to meet in a few days and complete their plans.

Ruth had a different idea. She thought Walter should take the job with Loutex, but that she should keep her job since his assignment would probably last only a few months in Snyder. Since they were living in a room in a private residence, she would not be alone. Walter was disappointed but could see the logic to her reasoning, so the matter was settled.

During most of August, Walter was busy making plans, studying the situation in Scurry County, familiarizing himself with Loutex and getting better acquainted with Thompson, Fogelson and the other associates in the company.

On the morning of August 30, 1922, he loaded his belongings into a Model-T Ford roadster, told Ruth good-bye, then went to the Loutex office in the Magnolia Building for a last briefing. Finally he shook hands with Thompson in front of the Adolphus Hotel and headed west. He traveled to Fort Worth and then over unpaved roads through Parker, Palo Pinto, Eastland, Callahan and Nolan counties, barely skirting Mitchell County just south of Scurry where the first West Texas oil had been discovered and into bone-dry Scurry County.

He spent the night in Sweetwater. He had a short trip left to get to Snyder, but he had pushed his car too hard the day before. Most of the second day was spent keeping the radiator cool. But on he went from Sweetwater where he veered north into Scurry County and on to Snyder in almost the center of the county. It was a grueling trip from Dallas and took 48 driving hours. His principal enemy along the way was the sandy roads which made killing demands on his T-model, probably the only car that could have stood the trip. But there were also blowing sands, driving dust and torrid heat to cope with. There were no roadside parks or cafes. He had to eat in grocery stores on a fare of crackers and cheese or baloney, washed down with Coca Colas or red soda pop. He promised himself he would never ask Ruth to make the trip in a Ford. He was glad she had stayed home.

He arrived in Snyder the afternoon of September 1, his wedding anniversary. E.J. Anderson, co-owner of the Snyder electric company, greeted Walter as he crawled out of his little black roadster. He was worn to a frazzle—his black hat, his clothes and his black moustache covered with West Texas dust. Anderson, not only the little town's utility magnate, but also its leading realtor, was delighted to see Walter. He welcomed him on behalf of the whole county, where the people were anxious to cooperate in any way to boost their own well-being and the county's economic welfare. Oil was being found in many other parts of Texas and they believed it would be found in their county.

Two months earlier Leon E. English, a respected hard-rock geologist, discovered evidence of a structure indicating the possibility of oil along the Colorado River, 15 miles south of Snyder near Ira. After this he made a reconnaissance check, spending several days measuring dips in the outcrops. On July 19, when he showed geologist W.A. Reiter over the area, English's findings were confirmed. Reiter was one of the most prominent geologists in the state.

Tommy Thompson had previously visited Snyder to look into lease prospects and found a warm welcome. On July 11 he had met with a group of farmers who said most Scurry County landowners would lease their land for oil. Nothing was said about lease prices or other details. With this encouragement, English, who had been engaged by Loutex as a company geologist, proceeded with plane-table work with an assistant named Kirk Ratliff. English had been joined about a week before Walter arrived by Bill and Nancy Nye, a husband and wife geological graduate team from the University of Oklahoma.

The day after Walter arrived, English had briefed him on the entire history of Scurry County and Loutex plans in the county. Two days later they met with "Cub" Murphy, a farmer in Ira, to arrange a meeting with as many farmers as possible. A large number attended the meeting the next day in the Masonic Hall over Taylor's store. They were enthusiastic about Loutex's interest in drilling.

Walter explained that Loutex was a new company and still poor but that it would be able to test Leon English's theory about oil in Scurry County. English spoke briefly to say geology indicated a nosing of the same type he and Reiter found in Mitchell County near Westbrook a few months earlier to open oil production in the Permian Basin.

Walter explained that Loutex was not able to pay anything for the leases, but if the company could get a block of 10,000 acres together, it would obligate itself to drill a well. He said materials could be brought in and all titles cured and other operational arrangements completed, but leases would have to be free with the guarantee of a test well on the block. The farmers agreed. They couldn't drill a well and they knew one well could trigger exploration for oil. Before the meeting was over, he had informal pledges for more than 10,000 acres.

The next week Thompson and Fogelson visited Walter and the geologists and were satisfied with the leasing campaign. Most of the farmers were signing up, but Walter had a big territory to cover. Reiter was back with the other geologists going over Leon's notes.

Late in September a test well on a Gulf Oil Corporation block southeast across the river from Ira in Mitchell County exploded with a wild roar blowing salt water and heavy air gas. Eventually the salt built up on the derrick timbers causing them to crumble. English and Nye got a sample of the gas and found it to be mostly nitrogen and a small amount of helium—very unusual, but no real indication of the presence of oil.

Meanwhile Walter was completing his leasing campaign which took several months. He had opened a Loutex office over the Snyder National Bank Building with E.J. Anderson, who often accompanied Walter on his visits to the landowners. Most titles to the land were in good condition because much of it had been in the same families for several generations. Some abstracts were only two or three pages long, going back to Mexican titles, but there were others with serious defects which had to be cured. There were a few stubborn landowners who wanted to be left with unleased tracts in the block that would make them rich if Loutex found oil. These windows had to be closed. It was an unexpected, time-consuming nuisance. Most of the 10,000-acre block he picked up was on the Ira structure, so named by English and Reiter. In later years it would be known as the Sharon Ridge where the prolific field was found at a depth much greater than 1923 equipment could penetrate. Most of the geologists working on the project for and with Loutex had been associated in the past with Julius Fohs, one of the most renowned earth scientists in the world.

It was late January before Walter got the necessary 10,000 acres bound into a single block. Along the way it took a lot of conversation. Most farmers just wanted to see someone and talk. Judge Buchanan, who owned the abstract company, made things as easy as possible. The county judge, Horace Holley, went with Walter on many of his trips, notarizing signatures for no remuneration. The two men were immediate friends.

By the time he had completed the leases, Walter had most of his equipment assembled and most of his crew hired. Eventually, however, he and Fogelson and Thompson had to man the rig because of the inexperience of some of the field hands.

In January it was obvious his work would be permanent, so Walter made arrangements for Ruth to give up her job and the room at Mrs. Pratt's and move to Snyder. Loutex realized Walter needed someone to help him with the paper work and Ruth was perfect. She could take dictation, type, keep books and do everything else necessary to relieve him for the outside work that was mounting daily. They even raised her salary to $150 a month. The $450 they were making together, plus all expenses, was only slightly short of fabulous for the times.

Walter had gone to Dallas by train so they could drive Ruth's car to Snyder for her use. He still had his field car. After storing their furniture, they drove to Snyder over the same horrible roads Walter had traveled the first time. Ruth's car was an Essex and considerably more comfortable than his Ford roadster.

They took a room in an old hotel built over a former warehouse. The dining room and other facilities were on the ground floor with the bedrooms on the second floor. Later they took two rooms in the more comfortable Manhattan Hotel across the courthouse square. After that they moved in with a private family named Smith until they bought a furnished cottage after their first year there.

In Snyder everyone knew Walter as "Bill" for some unexplained reason, except that his middle name was William, which he had never used before. Shortly after their marriage, he had started calling Ruth "Maggie" after the female character in the comic strip, "Bringing Up Father." The two names, Maggie and Bill, were to stick with them throughout their long stay in Snyder. He called Mrs. Lechner "Maggie" for the rest of their lives, but when he left Snyder, he left "Bill" behind.

Snyder was an interesting little county seat of slightly more than 2,000 inhabitants. It had been settled in 1877 by a man named Pete Snyder, who opened a trading post there. A community of buffalo-hide tents and dug-outs, it was first called "Robbers' Roost" because desperadoes and fugitives found a haven there.

Snyder laid out a townsite in 1882. Then in 1884 when Scurry County was laid out, it became the county seat and a post office was organized. By 1890, the year Walter was born, it had a population of 200, with three churches, a school and a weekly newspaper. In 1908 the Roscoe, Snyder and Pacific Railroad reached the community. Three years later the Panhandle and Santa Fe built a branch line through the town. In 1915 there were two newspapers, the *Signal* and the *Free Press*.

Scurry County was located at the foot of the Llano Estacado, or Staked Plains of Texas. It was about 900 square miles of prairie drained by the Colorado River and tributaries of the Brazos. Elevation was from 2,000 to 2,800 feet. Mesquite was the only timber available, but the soil contained magnesium, clays, gypsum, bentonite and coal.

Dairying was the chief source of farm income with beef cattle second. Farm products included cotton, grains, wheat, peanuts, peaches and some truck vegetables. Sheep and poultry were other products. In the early days of the county an ox trail between Snyder and Dallas provided a route for merchandise and supplies for buffalo hunters. By 1922 the county was, of course, largely agricultural and fairly prosperous.

Ruth arrived in Snyder about the time Walter was completing his block of 10,000 acres and getting ready to spud in the first well. Reiter staked the location on the basis of English's geology. The drill site was on the J.J. Moore farm. English was working for Reiter and Foster. All three had an interest in Loutex. Tommy Thompson, who had been a broker in Denver, persuaded Oaks and some of his Denver investors to put up money to help form Loutex. Sneed, an oilman from Tulsa, also had money in the corporation. They, along with Fogelson, Thompson and Lechner, had approved the location.

English had found the significant nosing coming out of the Westbrook field in Mitchell County. From there it extended through Borden County and into Garza County. Borden was west of Scurry and Garza was north, but the nosing went right through Scurry. The Nyes, Bill and Nancy, had joined their friend English and were helping him in any way they could.

Ruth Lechner drove the stake for the location of the first well in Scurry County. Later when the rig was in place, Mr. Moore, along with his sons, Pat and Mike, helped fire the boiler.

The Moore No. 1, which was to become the first oil well in fabulous Scurry County, was started on February 5, 1923. It was about a mile and a half south of Ira on the Sharon Ridge, which was named for the Sharon school nearby. Mr. Moore was given the honor of digging the first few shovelsful of dirt for the cellar. Pat Moore helped with a grubbing hoe. About 100 farmers and ranchers were present for the ceremonies. Ruth and Elizabeth Autry, a friend, accompanied Lechner and English to the spudding.

Walter had started keeping a diary during the war and had never stopped. After he was married, Ruth helped him, making entries herself on many occasions, especially during the long Scurry County episode.

The diary records numerous hardships and delays before the J.J. Moore No. 1 came in on October 9, 1923, as the first producing well in Scurry County.

Late in February, 1923, Leon English took one of the nation's most distinguished and influential petroleum geologists, Wallace Pratt of the Humble Oil & Refining Company, over the Sharon Ridge outcrop. Pratt agreed that Loutex had a favorable structure for its exploration but refused to recommend it to his company because there was not convincing evidence of commercial production in West Texas. He said he would hold off from getting involved in that area for the time being. This attitude probably cost the normally progressive company billions of dollars. But it was largely independents who were interested in such rank wildcats in those days, much to their own benefit. It was rare for a large, integrated company to make a significant discovery.

The oil industry in this country was relatively new in 1923. In 1922, for instance, the nation had its first annual production of a half billion barrels. Only seven years later, in 1929, about the time Walter was departing Scurry County, the production would reach a billion barrels for the first time.

The United States government had just started a long series of protests against its former allies, the British and French and Dutch, who had practically taken over the promising oil prospects in the Middle East without including the Americans in the concessions. Washington was bringing pressure on the oil companies to get into action in Saudi Arabia, Iran and Iraq or anywhere else in the world oil might be found. The word was that America might be out of oil in 5 to 10 years. Few of the oil companies cared to wander out of their own North American geological province, but they didn't want to go out of business and they didn't want to incur the ill will of the government. Walter watched this development with interest and some wonder. The independents believed they could find all the oil the country needed if they could afford the big risks involved.

Late in 1922 the famous Barroso No. 2 well had blown in on the edge of Lake Maracaibo in Venezuela and had built its flow up from 2,000 to 100,000 barrels per day. This was more threatening to the domestic U.S. oil wildcatters than the growing production in Mexico.

Late in May, 1923, the Santa Rita well ushered in the Big Lake field in Reagan County, south and slightly west of Scurry.

The first oil found on University of Texas lands, this was the beginning of one of the richest university funds in the nation. The field was an extension of the great Permian Basin area that had its beginning on Leon English's "nose" at Westbrook in Mitchell County on July 18,1920. Big Lake was also the first bonanza well in West Texas. Many more were to follow over the next half century and probably far beyond that. But the next five or six years would see West Texas bloom into the greatest producing area in the world at the time, with such magnificent finds as the shallow Yates field in Pecos County with one well that doubled the output of Spindletop's 100,000-barrels-per-day Lucas gusher.

The days came and went on the "pore-boy" well Walter was drilling near Ira. His job was varied and interesting. But the days were long and the work was often back-breaking and uncomfortable.

Snyder was not one of the world's beauty spots. Snow and sleet and ice storms dominated the winter. The summer was hot and sandy and there were dust storms, hailstorms and tornadoes. Along the Colorado River, especially in the southern part of the county, lay the shinnery sand, which was the "blow sand" and the fuel of the driving dust and sandstorms in the area. The storms came out of the hills and canyons along the Colorado and swept across Scurry and other counties in West Central Texas like a sandblast, peeling the paint off houses and cars, damaging buildings and glass on farms and in the towns, killing crops, blinding livestock and chasing people to shelter. Tornadoes were frequent killer visitors.

Fearsome "blue northers" frequently appeared on the distant horizon like eerie monsters. They were things of wonder and beauty until they unleashed their fury. The temperature would plummet and the wind and hail and whatever it contained would beat down on the countryside like a blast of a million times a million deadly machine gun pellets out of the dark blue sky.

Walter's work day was seldom less than 12 hours, often 18 and sometimes 36. Several times it ran 72 hours at a stretch. The old rig the contractor was using was almost always broken in some way. So were the field cars as they were worn down by the terrible dirt roads and the countryside that Walter often had to cross without roads. The workers were often difficult and on occasion Walter had to fire one. And good field workers were scarce.

Once a lawsuit filed by the drilling contractor shut off work for weeks. Finally Loutex bought the contractor's run-down rig and equipment. Walter was already driller, superintendent, manager and plain field hand, so he became contractor in name. Another of his duties was entertaining visiting firemen and acquainting them with Loutex operations.

Throughout all this he was covering the whole territory making new lease and drilling deals. The farmers lived miles apart and were lonely. Lechner spent many hours eating and drinking with and just plain listening to men who seldom saw a visitor from even the next farm or ranch. But it paid off. Walter made friends with 90 percent of the people in the county. His manner was like that of a good bedside physician. He got most of his leases simply by promising to drill a well on a block and offering the chance of wealth without cash investment for the landowner and prosperity for his county.

Ruth turned out to be one of the biggest $150-a-month bargains in petroleum history. She took all the paperwork out of Walter's hands and managed his home at the same time, keeping him happy, comfortable and well fed while he carried on his man-killing schedule. They had time occasionally for a social evening at the theater or in one of the comfortable cafes or hotel dining rooms in Snyder, or for an occasional trip out of town to Abilene, Sweetwater or some other nearby town. At times Walter tried golf to relieve the tensions of his work. Often he found time to work around the house, mostly when the rig was down for repairs. Once he took off to help a friend build a pony shed. Another time he built a windmill for another friend. Even on rest days he was a compulsive worker, building an outhouse, shoeing a horse or milking the cows for sick friends.

Drilling in Scurry County was accompanied by some puzzling geological phenomena. At one depth the crew ran across non-flammable gas so cold it could be used for refrigeration on the floors of the derricks. When a highly pressurized reservoir was hit one day, a strange blowout occurred that destroyed much valuable equipment.

The tool dresser was running the screw while the 10-inch bit was in the forge being heated to dress when the blow came. Before he could get down off the knowledge stool,[1] mud and water were blowing 50 to 60 feet into the air. Thinking it was gas, he grabbed the roustabout off the lazy bench and dragged him down the belt hall and out of the engine house. He shut off the steam under the boiler. They raced through the deep shinnery sand a safe distance. They expected a tremendous blast of flame since the forge fire was still burning.

After about 30 minutes and no fire or explosion, the tool dresser went back to the rig and cut off the forge fire. By that time the 10-inch bit had melted off. He went around the rig with the wind and could not smell gas. The wire line had been cut by the strong blow of cold air, rocks and sand. Destroyed parts had to be replaced and tools fished out of the hole.

The cold air, which was nitrogen and helium, was unique in the oil country at that time. After the accident Walter decided to send the gas through the pipes into the boiler and use it instead for drilling power. The air also kept water jugs ice cold, froze ice cream and iced down watermelons in just a few minutes. If left any longer, they would freeze.

That was the first of three troublesome blows on the well. The second was at the depth of 3,575 feet which forced the operators to abandon any hopes of going deeper. Then, when the well was brought in at the shallower production depth of 1,800 feet, it blew again, further delaying completion.

On that memorable October 9, 1923, when the famous Moore No. 1 started flowing oil, Walter was in charge. What a fine reward for months of coaxing a stubborn, cantankerous boring into one of nature's most precious treasures. The area around the well was crowded with business leaders, politicians, farmers and ranchers when the well was completed and turned into the tank.

---

[1]The knowledge stool or bench is a three-legged stool belonging to the driller. It sits next to the knowledge box in which the driller keeps his orders and reports.

It came in making about 30 barrels of 28-gravity oil a day. The oil was freed from some 230 million years of entrapment in the San Andres formation, 1,800 feet below the earth's surface. The well had gone down to twice the depth but had been plugged back. The flow was relatively light. It wasn't a gusher or a new Spindletop, but it was a fairly good well that would eventually produce a half million barrels of oil by 1973 on its golden anniversary, when it was still producing. It was prophetic for West Texas, one of the world's major oil and gas producing areas, since later great oil deposits in the San Andres would be found at about 5,100 feet in the prolific Midland County fields in extreme West Texas.

Everyone was delighted and plans were made to confirm the discovery by starting the J.J. Moore No. 2 not far away.

After that, the weeks and months and years seemed to skip by. Walter drilled five good, but small wells for Loutex. The old cable tool rigs were shifty and perverse. Someone got a leg broken or a finger cut off or was shaken up in some way almost weekly. Walter was busy searching for nitro with which to shoot the wells or locating some new tool or taking parts to a nearby town for repairs. He was constantly in communication with supply salesmen or gathering up new leases or entertaining some of his associates.

Once Tommy confided that he and his other partners were thinking about selling out and asked Walter's opinion. Walter seriously considered buying Loutex. Financing would have been simple for him since he was held in such high esteem by everyone in Snyder and most of the wealthy farmers and ranchers in the county. Ruth, however, advised him against it and he dropped the idea before mentioning it to anyone else.

On another occasion Loutex considered going into a Mexican deal. Probably had they realized that the boom would soon become so fabulous the Mexicans would eventually swim in oil, he and Tommy and Buddy would have given the idea more thought. Later when foreign holdings in Mexico were confiscated, they were glad they had not.

Once Walter and Tommy considered buying out their partners, but Ruth nixed that deal also.

Walter fought rain and hail and mud constantly in the winter months. Muddy roads and broken-down cars were two of his nemeses. One day he had four flat tires from Snyder to one of his wells.

As the years passed, Walter was often sick or injured and sometimes his injuries and illnesses were serious. At other times in the winter they were merely nagging colds or smashed fingers or cuts. His stomach acted up due to the fast eating in roadside greasy spoons.

Walter was doing most of the work on Loutex operations. Occasionally Buddy or Tommy would come out and put in a day's work on the rig, but that was not very often and Walter never knew when it would happen. They were raising money or making contacts most of the time.

Therefore, when Al and I.B. Humphreys, sons of wildcat oil promoter Col. A.E. Humphreys, bought out Loutex, Walter was agreeable. He got his same interest in the new company (to be known as Northwest Oil Company), the same title—superintendent and tool pusher, with the tools belonging to the Northwest Co., which meant all the work in the field—and the same salaries for himself and Ruth as he had in Loutex.

But things didn't get much better. In fact, in some ways they got worse. But Walter was a lover of adversity and hard work. There wasn't anything he couldn't do on a rig and his knowledge of the geology of the area increased by the day. He was a fine mechanic and knew the land and all there was to know about a farm or ranch. To him the people of Snyder and Scurry County were the salt of the earth and the feeling was mutual.

Ruth and Walter still managed to go to Sweetwater, Abilene, Colorado City, Westbrook or even Ranger for a dance, a theater performance or dinner. Once or twice he returned to Snyder at 2 or 3 in the morning from such a night out and went directly to a well in trouble and worked until noon.

He was always so tired he didn't even notice it. Still in his mid-thirties, he seemed to be getting tougher and building up more endurance by the day.

With Northwest he completed several good wells, but his relationship was not the same as it had been with his friends in Loutex. Sometimes he almost felt like an outsider.

During this time, Walter negotiated with the Packard Motor Company to open a dealership in Abilene. It was the only agency in West Texas. In fact, the vice president of sales for Packard sent his brother-in-law to run the agency for Walter. The venture was never a spectacular success, but it did keep the Lechners and their associates on good, inexpensive wheels. It was the only time he ever ventured into a sideline outside oil, even though this could be considered an adjunct to the oil business. The experience taught him to stick to his last, which was oil.

One day Walter and Ruth decided it was time to give up their life of excessive work, small salary and few dividends. He was ready to do something more on his own and see if he couldn't find that legendary pot of gold at the end of every wildcatter's rainbow.

The Packard deal turned out to be an asset after all. The dealership had accumulated about 30 used cars. He sold them all for what he could get, which turned out to be about $15,000, and left Northwest Company.

Before he left, he had managed a well in the Wortham boom for Northwest. Not only had he succeeded in his leasing and drilling program, but he met Beauford Jester from Corsicana, a young lawyer who, like Allred in Burkburnett, was working on titles. They became fast friends.

Another friend of Walter's was Johnny Farrell, a lease man for Marland Oil Company. The Farrells and the Lechners had become friends in Scurry and surrounding counties where both were gathering leases. They had spent many evenings together at dinners and shows and dances in Snyder and surrounding towns. As in the case of Jester, Farrell would become significant to Walter's future.

Johnny Farrell was red-haired and about 5 feet, 10 inches tall. He was of medium build, rugged, and weighed around 170 pounds. Clean-cut and a gentleman, he was a former Canadian with a fine manner that enabled him to do his job as a lease and land man most effectively. Farrell was making his home at the time in Abilene. Once he engaged Walter in a leasing swing through Howard, Mitchell, Borden and Garza counties. During the foray, the two men became even better acquainted and learned to respect each other's talents in the leasing business.

Fate would later throw them together in the most important venture of their lives.

When Walter finally left the Northwest Company, he made an agreement with D. Harold Byrd, a young geologist who held a world record of 56 consecutive dry holes and had become known as "Dry Hole" Byrd. They started a well in Fluvanna with Henry A. Harmon of San Antonio as a joint venture. Walter had adopted a policy early in life of never entering into a pure partnership. He was a complete independent.

The project for Byrd, Lechner and Harmon was a well on the Koonsman lease for Edward L. Doheny, probably as well-known for his role as a defendant in the Teapot Dome scandal as for his achievements before and after in petroleum exploration. While this well was in operation at Fluvanna in northwest Scurry County, Walter and Ruth continued to live in their cottage in Snyder. For eight months the cable tools tried unsuccessfully to reach the deep oil treasures of the county. Walter had suggested the use of rotary equipment, but his associates still believed rotary bits could not penetrate the red beds of central West Texas. The well was finally abandoned as a dry hole.

The men moved to the Monroe lease in Ward County, where Walter sold his tools and interest to Byrd for $3,000.

The Lechners sold their cottage for $4,500, said their farewells and headed for Dallas. When they reached Dallas, Walter was told by Jimmy Nowlin, Ruth's brother and a scout for Atlantic, about a big lease play by Gulf Oil Corporation in Grimes County, north of Houston and near his old alma mater, Texas A&M College. Walter went to Grimes County and formed a joint venture with G.C. Tisdale. The county had indications of oil and the leasing play was little short of sensational. Both Walter and Tisdale spent their time quite profitably taking leases for Gulf for 50 cents per acre. Grimes was one of the counties in Texas with no oil production. In fact, it remained dry until 1951 when Gulf discovered a small field.

In 1929 the future looked good for a man as competent, industrious and aggressive as Walter. That is, until that fateful day in October when the bottom fell out of the stock market and about 16 million shares were dumped on the market for any price they could bring and the market slumped more than 200 points. The Great Depression was on!

About the same time, Ruth informed Walter she had decided to run a home and leave the work for him.

So the end of the decade saw the end of an era in Walter's career and the beginning of another.

Chapter 6

From Desdemona to Scurry County and then to Grimes and Borden, Walter Lechner had learned almost as much as one man could learn about the oil fields. He knew the art of leasing land and how to put a deal together. He had a good understanding of geology and petroleum and knew how to drill a well and what to do with the oil when he found it.

By 1930, he was ready to do something on his own. He had learned an important lesson—avoid partnerships at all costs. He was ready to become a wildcatter. Always before him, as it was before every wildcatter in the oil business, was that pot of gold at the end of the rainbow—that one strike that would pay for all the hard work and frustration and failures he had experienced. It was about time Walter got on with his rainbow chasing. By mid-1930 he was approaching his fortieth birthday.

Some young wildcatters have hit their bonanza on the first well; others waited until they were well into their seventies, like Dad Joiner. Most of them, he fully realized, never hit anything that would put them on Easy Street. The odds were terrible, but the opportunities were boundless.

If he ever wondered why he was impelled to remain in this risky racket, he made no outward sign of that feeling. Few men who get into oil ever get out. That is, unless they are absolutely forced out by economic conditions or health.

Wildcatters are an especially determined sort. When once they join the battle with mother earth to extract the most precious of her riches, apparently they cannot simply quit, no matter what comes, as long as the economic breath of life remains in them. Most wildcatters would rather be top water minnows in oil, with an almost imperceptible hope of some miracle converting them into a whale some day, than be a banker, a college president, a merchant prince, an industrial tycoon, the publisher of a great daily newspaper or anything else that marks a man a success.

Walter had long since despaired of being a court reporter, or a textile engineer, or an automobile magnate, or a railroad baron, or a refiner or a financier. The day he walked onto the sod at Moonshine Hill at Humble he was trapped. He didn't know it until he started to leave France after the war, but he was just plain hooked by the smell of crude, the noise of a rig, the handshake of a landowner and the art of putting together a deal or making a trade.

When he and Ruth returned to Dallas in 1930, the depression which had started with the market crash on Black Tuesday, October 29, 1929, was widespread. Oil was almost the only business that had not been ravaged. Even so, many in the industry had been wiped out because their old investors were often among those in the breadlines, especially where the money had been— mostly in the Chicago, New York and Boston areas. Economically, the nation and most of the world was in panic. Walter and Ruth were still in the small duplex apartment owned by Walter's Grimes County friend, G.C. Tisdale.

Walter heard of some oil activity in Miller County, Arkansas, and decided to investigate. After paying $100 down on a Model-A Ford, he left Ruth to take care of their home and set out for Arkansas. On the way he passed through Rusk County to look at a wildcat being drilled west of Henderson, Texas. When he got to the site, there was no activity, not a person near the location. Dad Joiner, the venerable wildcatter whose name was soon to become a household word across the oil country, was not even there. The little tea kettle rig on the Daisy Bradford lease was shut down for repairs. So he moved on.

By the time he reached Miller County in Arkansas, the lease prices had jumped to between $2,500 to $5,000 an acre. Walter couldn't lease one acre at those prices, so he turned around and headed back home. Fate or coincidence or the desire to visit some of his friends of the early Texas and Gulf Railroad days enticed him to Longview where he took a room in the comfortable Gregg Hotel. He was conscious again of the haunting memory of Dr. Tucker's prophecy.

The first friend he thought of was Dr. W.D. Northcutt, so he ambled toward the Terrell Drug Store over which the doctor had his office. Dr. Northcutt was delighted to see him. They went into the drug store and ordered coffee as they talked about the seven or eight years since they had seen one another.

Walter told Dr. Northcutt of his bone-breaking and frustrating days in the university of oil in Burkburnett, Scurry County and elsewhere. The doctor asked Walter why he didn't stay in Longview and drill a well and make his old friends rich.

"I'd do that, Doc, if I could get sufficient acreage together to warrant the drilling of a deep test."

Northcutt replied that he knew two men who might help. One was Diamond Joe Reynolds, who had a block of land north of Longview. The other was B.A. Skipper, who had a block west of town. In fact, he said, he had just been talking to Reynolds down the street. Northcutt left and returned a few minutes later with Reynolds, who said he had about 4,500 acres in a solid block. He wanted $10 an acre for a lease.

"That's not what I want," Lechner replied. "I want acreage free of bonus so that money can be spent drilling a well."

Reynolds was not interested. When he left, Dr. Northcutt called Skipper. They agreed to meet the next morning at the Gregg Hotel.

Skipper arrived at 9 o'clock with a map. It did not show any ownership, but it seemed to cover about 2,000 or 3,000 acres. The two men had an understanding and respect for each other from the outset. Walter said he would have to have the leases free and the block should be built up to about 10,000 acres to make it attractive to an independent operator or some oil company. Skipper said he thought that could be done and added that the farmers were giving him the leases free.

"That's fine," Lechner said, "and I would expect you to give them to me free." Skipper said that was the way it would be. He agreed to help assemble the block and put it all in Walter's name. They settled on a 50-50 split of all the leases they could assemble, including Skipper's 2,300 acres. A simple handshake in April, 1930, sealed the understanding between Lechner and Barney Skipper. New leases were taken in May, June and July.

Only one well was drilling anywhere near Skipper's block at that time and that was almost 30 miles to the south in Rusk

County where Columbus Marion Joiner, better known later as "Dad," was engaged in the "pore boy" operation on Daisy Bradford's land west of Henderson. It was the wildcat Walter had visited on his way to Arkansas. Everyone from the biggest major to the poorest independent had shrugged it off as a doomed wildcat since two previous tries on the same lease had been abandoned as mechanical failures. Others were skeptical because of the antiquated rig and rusty drill pipe they were using and the scarcity of funds which necessitated frequent shutdowns while the native crewmen could get back to their own farms to help with crops—and to eat.

Also no one believed the self-styled geologist, Doc Lloyd, when he reported that oil would be found there in the Woodbine formation at 3,500 feet and that a field 40 miles long and 12 miles wide would result. Geophysics had proved beyond doubt that, any geological guesses to the contrary, there was no structure to trap a migration of oil in Rusk or Gregg counties. Most respected geologists had condemned the area for oil. Only a few had not. Among them were Dr. Hugh Tucker and E.A. Wendlandt of Humble.

So when Walter Lechner and B.A. Skipper agreed to take Skipper's leases and try to build them up to 10,000 acres on a 50-50 basis with Lechner taking the lead and Skipper supplying the knowledge of the area and its people in rounding up more leases, Lechner was taking on an almost impossible task of promoting a well west of Longview, in Gregg County, Texas.

He had a valuable partner in Barney Skipper. A native of Gregg County, Barney had been actively promoting oil since 1911 in the hope of pulling the dirt farmers and other people in the triangle between Tyler, Longview and Kilgore out of poverty. His father had believed oil would be found in Gregg County. When Skipper first started taking leases, only one oil lease had been made in the history of the county. Shell McLaughlin had leased some land in 1906, but did not renew it and did not start a well.

Barney Skipper made some of his first money as a result of oil. In 1901 when the famous Lucas Gusher blew in at Spindletop, south of Beaumont, to usher in the fuel oil age, Barney got a job as a butcher boy on one of the railroads that provided excursions to the field. It was a profitable and educational experience, and he never forgot how poor men got rich in oil and how Beaumont and Port Arthur, and even Houston, prospered from the great field in a very short time.

His father often said, "If oil isn't found in Gregg County in my lifetime, my boy will find it." Then he would pat Barney on the head and say, "Ain't that right, boy?" and Barney would smile and nod in vigorous agreement.

Barney Skipper went to school only a few months and never really had time to learn to read or write, although he developed these skills on his own as he grew into manhood. He was a natural salesman and started his career working in men's clothing stores in Dallas, San Antonio, Little Rock and Birmingham.

In San Antonio in 1908, while he was working for Joske's Department Store, he met and married Mary Obadam who, born in Budapest, was the daughter of an immigrant family.

Later he moved to Dallas and worked in a store on Commerce Street, ironically only a block from the Roberts Motor Car Company, on the site where the Adolphus Hotel was later built, where Walter Lechner was working at the same time. They did not meet then, however. Skipper worked there four months. He recalled the time because when he moved to Dallas, his only child, a son also named Barney, was four months old and when he left, the child was eight months old.

One day he just got tired of the big city and decided to go home to Longview. He did not even mention his decision to Mary, telling her only that they were going to Longview on a visit. He had a Hupmobile and a few hundred dollars in his pocket. He paid down on a small $300 shack on the road between Longview and Kilgore. After a week, he told Mary, "We are now living in Longview." She was only slightly surprised. He had obtained employment at Perkins Brothers store in Longview for $15 a week, considerably below even the $100 a month he had made in San Antonio. His boss in Dallas was shocked at his leaving since he had paid Barney more than he had been making in San Antonio to get him to Dallas and had handed him a nice raise a week before he quit.

"Why are you quitting?" his boss asked. "I just gave you a good raise."

"I'm going into business for myself," Barney replied, "even if I have to sell peanuts."

After he had been back in Longview for about six months, Barney came home and told Mary he had quit his job at Perkins Brothers. For a moment Mary was frightened, but soon her confidence in her determined husband overcame her fears. He said he was never going to work for anyone else as long as he lived.

He had chosen real estate and had already made arrangements to promote and sell lots in a new residential development in Longview. Then he opened an office in the Trice Building for which he paid $10 a month. The first month he made $100 which brought him back to the income he was making at Joske's.

From the day Barney landed back in Longview in 1911, he started talking oil. He buttonholed everyone he could stop to tell them there was oil under that soil west of town. He would pick up a handful of dirt in the development where he was selling and show his customers how different it was from the soil where there was no oil.

He talked oil so much he actually scared some of his customers away and there were times when he and Mary and little Barney didn't have sufficient food. But he would turn to Mary when things were toughest and tell her, "We're going to root hog or die."

Barney learned how to take leases and several times he went to his friends and neighbors in the county and blocked up large areas of farm land, only to lose some of it later. He always told those who gave him the leases free that if he could not get a well drilled they could have the leases back whenever they wanted them.

Once he went to the leading businessmen of Longview and asked them to put up a pot of $5,000 to attract oilmen. The only person who showed interest was Dr. Northcutt, who often helped him promote his idea. No one else would listen. For a while in about 1918, he worked for Colonel Featherstone, the railroad promoter who had built the line from Beaumont to Port Bolivar. He was then promoting the line from Longview to Ore City to become part of the Santa Fe's system that would provide an outlet for East Texas iron ore. At that time iron ore seemed a far more likely prospect for East Texas progress than oil. Again

Skipper and Lechner, who worked for the same boss, were thrown together but did not become acquainted.

Skipper's one-man promotion of oil never met with much enthusiasm. People, tiring of his constant talk about oil, said he was crazy. Because others were so pessimistic about oil possibilities, Barney was able to apply his real estate profits in royalty at bargain rates of from 50 cents to a dollar an acre all over Gregg and Rusk counties.

In 1928, on borrowed money, Skipper wrote letters to about 750 major companies and independent oilmen offering them his block of leases free if they would drill a well. They were told that one well would validate the entire block by the terms of the individual leases. He and Mary and other members of the family, including young Barney, helped to get the letters typed, duplicated, stuffed into envelopes, stamped and mailed. They waited for days, then weeks and months. There was only one reply.

Skipper could have blocked as much acreage as anyone required. The farmers and other people of northern Gregg County were living in a local, private depression that had gone on for years before the Great Depression. Their one hope was oil. Yet all the experts said lack of geological structure made finding oil in East Texas impossible. Geologists and geophysicists and even creekologists had condemned the whole strip of East Texas from Nacogdoches County north to Upshur, through Rusk, Smith and Gregg counties—almost everyone, that is, except Doc Lloyd, Dad Joiner and the poor people.

Old Man Joiner had come into Rusk County in the middle twenties and had started promoting a well. He had a geological report from his promoting crony, Doc Lloyd, stating that an ocean of oil would be found in East Texas in the Woodbine sand at a depth of around 3,500 feet. The experts denied this, saying the Woodbine would be much deeper and at least 1,500 feet below the salt water level.

A few geologists, even some from major companies, believed there might be some oil in Gregg and Rusk counties, but their number was small and their voices were drowned out by the more positive pessimists.

The stock market crash in 1929 and the Great Depression of the thirties did not discourage Barney Skipper. He had dropped most of his leases, which in 1928 amounted to almost 10,000 acres, but he still had about 2,300 acres left and knew that any good prospect to drill would permit him to build up a block to the specifications of the wildcatter willing to gamble on Gregg County.

There was still no sign of oil in East Texas. The Van field in Van Zandt County about 60 miles west of Longview was the nearest. The Joiner operation in Rusk County west of Henderson was struggling along, still about six months from its first show of oil.

It was at this point in April, 1930, that Lechner, at the suggestion of their mutual friend, Dr. Northcutt, met Skipper.

They talked, agreed to build up the block to the 10,000 acres Lechner believed necessary to attract an operator to drill a test well. They shook hands on it. Skipper's 2,300 acres were not all in a solid block, but he believed he could go back to his friends and fill in the vacant spots and add the acres Lechner believed they needed.

There was an immediate feeling of confidence between Lechner and Skipper. Lechner was a husky, 180-pound, cigar-smoking, anti-prohibitionist, constitutional Democrat who still sported that dapper moustache he had picked up in France during the war. He was a rugged man, proud of his rough hands resulting from 10 years of labor in the oil fields. He had acquired little more than experience in his lifetime of fighting nature for its greatest bounty. His bank account was down about as low as it had been since he left Texaco back in 1917. But he knew the oil game, had ambition and believed in that idea he got from his old friend, Dr. Hugh Tucker, over in Harrison County.

Lechner and Skipper had some definite dissimilarities. In the first place, Skipper had missed his chance for an education when he went to work as a boy to help his family. He was also a tee-totaler, regarding alcohol and tobacco as tools of the devil. He was about six feet tall, also rugged, and he weighed about 210 pounds. He was a devout Protestant and was devoted to his wife and their strapping son, Barney.

Like Lechner, Skipper was a man of vision, determination and ambition. He was a one-man chamber of commerce for Longview and a natural promoter. Skipper did not doubt there was plenty of oil west of Longview. All he was looking for was someone else with the good sense to recognize that simple fact.

In Walter Lechner he finally had his man. Skipper found a notary public named Tom Garner to work with Lechner. Walter told Skipper that, to simplify things since they were operating on a 50-50 basis, all leases would be taken in Skipper's name. This arrangement would also make it easy for Lechner since everyone he would be talking with knew B.A. Skipper and knew about his faith in oil. That faith gave them the same hope in Gregg County that the poor dirt farmers and little merchants in Rusk County had in Dad Joiner. Lechner and Skipper formalized their hand shake with a written agreement on July 19, 1930.

Lechner paid Garner his notary fee mostly with money he borrowed from Ruth or some of his friends, including Bob Thornton, president of the Mercantile National Bank at Dallas. The response to his campaign was excellent. Within four or five months he had increased the lease spread to 9,300 acres. A few windows remained in the block, but not many. Few if any others were interested in this area, so Walter had no competition. The only cost of the leases was the obligation to see that a well was drilled somewhere in the block. While the leases were all taken in Skipper's name, it was with the understanding that if Lechner got someone to drill, they would be transferred to his name.

Day after day, Walter drove through the beautiful rolling hills and wooded valleys of Gregg County and came back each night to the Gregg Hotel with a sense of achievement. Occasionally he would go back to Dallas to the duplex on McKinney.

Ruth spent most of her time at home. Sometimes, however, she would trudge the woods and farms of Gregg County. Once, when the weather was cold and wet and the streams were flooded, they went to see a landowner about getting a lease. Tom Garner was with them. They came to a swollen creek and knew it was impossible to drive the car across to the house. They decided to wade, but Ruth chose to stay in the car, a Model-A convertible coupe. The men intended to be gone only 30 minutes since the sun was getting low, but Walter's salesmanship took longer than anticipated and Ruth was left in the car for about two hours, half of that time in the dark. Her mood was not cheerful when Walter returned, but he had the lease. After venting her feelings, she succumbed to his happiness over the acquisition and they both laughed over the incident.

During most of this time Skipper was taking leases along the banks of the Sabine River for Humble Oil and Refining Company, all the while pleading with the company to move over into the area northwest of Longview where he said prospects were better. But they trusted their own scientists more than an uneducated promoter with a hunch. Once in a while Walter Lechner would contact the Humble people and try to succeed where Skipper failed, but to no avail. Humble had leased 900 acres of Skipper's own land east of Longview, even though he insisted it wasn't oil land.

Walter went to Dallas frequently to see some of the other majors and various independent oilmen, such as his friends, Thompson and Fogelson, who had been successful in recent operations. He visited the Sun Oil Company, Atlantic, Magnolia and every independent he knew in Dallas. He even went over to Fort Worth where there was a whole army of successful independents and offices for other majors operating in West Texas, including Gulf Oil, but they all gave him a courteous reception and a polite but firm turndown.

One man had the same idea Skipper and Lechner had. He was R.B. Whitehead, Atlantic's Dallas geologist. Whitehead was a colorful man, rugged and clever. He chewed cigars constantly until he had ground his teeth almost down to the gums. He submitted Walter's block to his district bosses, but they turned it down. In desperation, Whitehead went over their heads, to Philadelphia, the company's headquarters, recommending the deal based on his experience but without sufficient solid geological evidence to back up his plea. Not only was he turned down, but he was called into the Dallas office and reprimanded, even threatened with discharge unless he stopped wasting his time dreaming about oil in Gregg County.

Whitehead was furious, but was forced to accept their decision. He went back to Walter with the sad news and with the advice to stay with his prospect.

"Man," Whitehead told Walter, "you are sitting on a powder keg that is going to blow this place wide open if you can hold on."

One day Walter remembered a friend, John E. Farrell, who had worked closely with him. Frequently Farrell and his wife had played bridge or gone to shows with Walter and Ruth in Snyder. He was with Marland Oil Company in Scurry County then, but was now an independent operator in Fort Worth. Lechner did not know whether Farrell could handle the deal financially, but he decided to try him. A few enticing rumors were circulating about the Joiner well on the Daisy Bradford lease in Rusk County and that might help. The rumors were based on a report that an oil-saturated core had been accidentally left on the derrick floor of the wildcat. But big company scouts said it was evident Joiner's driller, Ed Laster, had tried to con them with a core stolen from a well in Van field, left conveniently on the derrick floor for them to find, which they did. They had even taken the core to a laboratory where the chemist assured them that it was from Van. It was a laughing matter.

Once in a while Dad Joiner, who had an office in Dallas also, would visit Lechner, and the two would commiserate with each other. Dad, always optimistic about his deal, told Lechner there were no games being played with the cores, but would not admit he had found anything. A few farmers were positive the Joiner well was running exactly as Doc Lloyd said it would run, geologically. They were guessing, but the word was enough to stir up minor interest. Oilmen never totally dismiss even the most obvious over-optimism about any wildcat well.

So what if all the good reports turned out to be even better than the hopeful farmers and merchants said? The Joiner well was almost 30 miles south of the Skipper-Lechner block. No oil field could be more than a few miles long. Walter had little hope when he finally went to see Farrell. Already the Humble and Shell tests east of Longview were virtually public information and discouraging.

Farrell was glad to see Walter and the two talked for some time about the good old days in Scurry County. Only a year or so ago both had left Snyder when the Sharon Ridge play cooled off and the Humphreys brothers deserted the area along with all except the most hard-headed wildcatters who could still get a stake.

Walter produced his map and showed Farrell what he was peddling. It was a nice map, but when Farrell asked about the nearest production, Walter only smiled and stuck his chubby finger on Van, 60 miles west of Longview, almost halfway to Dallas.

Asked what made him think there was oil in Gregg County, Walter told Farrell about Dr. Tucker's prediction and the hopes and dreams of Barney Skipper. Also, Walter smiled, the grapevine had it that Farrell might even have about $17,000 and that could drill a well. "And it must be true," Walter said, "because you haven't refused me on the spot like everyone else." Farrell laughed and invited him to dinner. Walter was happy to go and renew his acquaintance with Mrs. Farrell. It was the first thing he had heard in five months besides a cold turndown.

They talked some more and agreed to get together again in a few days. Farrell wanted to think over the matter. He said he had a partner, who might even have a partner or two himself. Walter said he could have all the partners he wanted. In fact, he added, he too had a partner, Barney Skipper, the man with the dream. A fortune teller once told Barney he was going to find oil in Gregg County, he said, as they laughed over a bourbon.

"Johnny," Lechner said, "I don't have any geology, just a solid block." Farrell said he understood. He also said he had been the recipient of one of Skipper's 750 letters and had thrown it away.

Farrell said he was interested and would like to know if Walter and Barney could deliver 5,000 acres for 50 cents an acre. Walter said that was possible, if he could select the location for the test well. Farrell agreed and suggested he return with Skipper in a few days when he could introduce them to his partner, W.A. Moncrief.

That was all Walter and Barney wanted to hear. This was the culmination of the endless toil of building up a 9,300-acre block and looking for an operator able to drill. Five thousand acres would leave him and Skipper with 4,300 free acres in the block, plus $2,500 to work on.

Chapter 7

Walter's trip back to Dallas after his meeting with Johnny Farrell was a happy one. He went directly home to Ruth and told her that maybe the struggle was over and that he was on his way to genuine success in the oil business.

The words of Dr. Tucker had come to Walter's mind many times during the months he spent with Barney renewing leases, bringing old ones up to date and trying to fill in as many gaps in the lease block as possible.

It will be recalled that in 1919 Dr. Tucker had picked up a clod of lignite in the woods south of Marshall and said that some day one of the greatest oil fields in the world would be found less than a hundred miles west. The land Walter and Barney had blocked into an attractive package for exploration qualified almost precisely.

As soon as Walter let Ruth in on the good news, he called Barney. The two dreamers talked so long on the telephone, both got worried about the cost.

The next day Walter was in Longview to meet Barney. He told him all about his conversation with his old friend, Johnny, and that it seemed to him the deal was in the bag. He said there were no details of an agreement, but that they had better start thinking about some because they were to meet with Johnny and a partner the next week.

They had 9,300 acres, so they agreed they could accommodate Farrell's requirement of 5,000 acres and still have a sizeable block for themselves, plus whatever else they could round up. Johnny had said they could give about 50 cents an acre, which would pay for some of the funds they had expended to acquire the acreage. They had paid nothing for the leases, but had been out for expenses, including travel and notarizing.

In his conversation with Farrell, Walter learned for the first time about the 750 letters Barney had sent out to oil companies, independents and brokers several months before the two men met. Farrell at that time had been offered a sizeable block of leases by Barney for nothing. In the meantime he had visited Gregg County several times to look over the ground. Then he had talked with geologists and other experts, but had received no encouragement.

It was the July core and the recent drill stem test on the Joiner well 26 miles south of Longview that had piqued Farrell's interest this time. As a matter of fact, it was the test that had sent Walter scampering back to his friend.

It was September 11 when Walter and Barney met in Moncrief's office in Fort Worth. In addition to Moncrief, others there included Eddie Showers and Frank Foster, who along with W.S. Noble, would help Moncrief raise the money to join Farrell in the rank wildcat venture. All the partners, including Farrell, were former Marland Oil Company men.

The contract meeting was rather uneventful, except for a slight disagreement between Walter and Moncrief. Moncrief had insisted on a clause in the contract stating that if Barney and Walter failed to deliver the acreage, the trade was null and void. This was reasonable enough, but Walter then insisted on a provision that if the money were not paid on time or if Farrell and Moncrief did not perform in accordance with the contract that the agreement could be nullified. Moncrief had agreed to this, but with some bitterness, which included the side remark that the deal was no good anyway.

When the session ended, however, all parties seemed happy and it was understood that the project would get moving as soon as possible.

Dad Joiner didn't really have much reason for his determined efforts to drill a well on Mrs. Bradford's farm. He had a friend named A.D. "Doc" Lloyd, who was a self-taught geologist among other things. Lloyd once drew a series of lines connecting all the big oil fields in the United States. The lines crisscrossed several times. At each crossing he established an apex. Then he drew lines between the apexes. These lines crossed in the vicinity of the Bradford farm. He called it oil's apex of apexes and said a giant field would be found there.

Dad's wife ran a boarding house in Ardmore, Oklahoma. Among the boarders was a Miss Springer, secretary for W. Dow Hamm, a bright young geologist for Shell Oil Company. Joiner asked her to take the map of apexes to Mr. Hamm and ask what he thought of the idea. What Mr. Hamm thought was that his secretary was trying to catch him on something. He kept the sketch several days, then gave it back to Miss Springer, who gave it back to Joiner. She told Joiner Mr. Hamm had said nothing at all.

"He didn't condemn it?" he asked.

When she said, "No," he probably took that as an indication Hamm considered it a good idea. Joiner packed his bags, said good-bye to his wife and children and caught a train for East Texas. Somehow he landed in Henderson. That was in 1927. He had been to East Texas frequently on leasing trips between 1921 and 1925 with several other Oklahoma oilmen, lured there by the idea the East Texas basin offered excellent prospects for wildcatters. But in 1927 he was on his own.

Columbus Marion Joiner was not an ordinary man. His life story reads like fiction. He had only a few weeks of schooling, but learned to read and write at home from the Bible. His first writing assignment was to copy Genesis in his aunt's home in Alabama. He learned quickly, soon became a merchant, then a state legislator, again a merchant and then a wanderer. His wanderings eventually led him to Oklahoma, where his sister was married to a Choctaw and was living with the tribe. He became manager of the tribe's oil leases. Later, as a wildcatter, he missed discovering the great Seminole field by only about 200 feet and missed finding the Cement field by even less than that. He was one of thousands of wildcatters who always come close to success but never quite make it.

It was there that he met Doc Lloyd, who succeeded in finding oil despite his lack of a degree. Lloyd changed his name from Joseph Idelbert Durham for his own reasons. He had a varied background as a druggist, a medical student, a medicine show promoter, a mining engineer and finally a geologist. He was also one of the great promoters of all time. He stood over 6 feet tall and weighed well over 300 pounds.

Joiner got Lloyd to fix up a prospectus which included a statement that oil would be found on the Daisy Bradford farm at around 3,500 feet. Later he got even more specific. It was this report that Dad used to sell leases or certificates in a 380-acre lease block mostly to Dallas widows.

He opened a tiny office in the Praetorian Building in Dallas, not far from the Kirby Building where Walter and Ray Hubbard would soon have an office. He would rifle through the newspaper obituaries, select the wealthy men who passed on, make a note of their addresses and later call on their widows with a great opportunity to invest in a prospect of high financial potential. It was a slow, tedious way of promoting a wildcat, but Joiner managed to eke out a living, keep his Dallas office and live in Overton with Walter and Leota Tucker and their family.

Tucker was a small-town grocer, a handsome, tall man about 40 years old. His wife, Leota, was attractive, energetic and intelligent. They were a great help to Joiner in his venture simply because they liked the old man. Dad, in 1927, was 67 years old, some six years younger than Lloyd, his geologist. The Tuckers gave Dad a room, board and space in a storage shed for an office. This was an unlikely cast of characters to drill an oil well that would bring in the world's greatest oil field and change the whole concept of the oil-producing industry.

Joiner, strapped for money, had started two tests prior to the Bradford No. 3. Both were mechanical failures. Most of his equipment was cheap, rusty and old, all Dad could rustle up with the low funds he was able to raise. The No. 3 well had been no picnic. Without an indomitable driller named Ed Laster, he never would have made it.

The oil scouts for the major oil companies had the well location on their routes as a place to go every now and then to pass the time of day. They had it at the bottom of their lists of wells to be watched.

September 5, 1930, was a day to be remembered. A man from Arkansas, H.L. Hunt, had come in with M.M. and Charles Miller, who had invented a new testing tool. They had offered to let Dad use it on his well as a sort of demonstration. Hunt was an oil operator from El Dorado who got his start there running a small card parlor for non-professional gamblers.

The first core of the well had been taken a month and a half earlier on July 21. From that date on, Laster, Joiner, Mrs. Bradford, her brothers and Doc Lloyd knew Laster's drill bit had penetrated oil sand. The big company scouts, one of whom had stolen a part of the core, believed it was a planted lure to get big companies excited so Dad could sell off his worthless leases.

But on this September morning there were no great secrets. Most companies knew this was a drill stem test. If it showed oil, chances were good a well would result after a few weeks. The results that morning were good—so good, in fact, that on September 22 there was a Joiner Jubilee at Overton where the farmers, landowners, merchants and other citizens whipped up a real celebration for the success of old Dad Joiner. Yet there was no well and no one knew when there would be one or how good it would be.

East Texas was hungry for something good to happen. It had suffered a depression for 20 years or longer, not just one year since October 29, 1929, when the stock market crashed. It was a country of lush growth, nice little farms, beautiful pine, oak, sweet gum and bois d'arc trees, rolling hills, clear streams and lakes. But its people had grown weary of toil and discouraged by poor crops, drought and no progress. They had heard stories of oil from Barney Skipper and his daddy before him as well as from Malcom Crim of Kilgore who had once blocked up 20,000 acres and offered it to anyone who would promise to drill just one well.

These things had inspired the people from time to time. But nothing had inspired them more than the little 70-year-old optimist in the pork pie hat, with the humped shoulders and quizzical smile. Now he was going to get himself a celebration, come oil or more depression.

Roxanna Petroleum Company, the drilling and exploration subsidiary of Shell Oil Company, missed by a little more than a mile finding the great blanket of oil as early as 1915. The Snowden Oil Company of Indiana had employed Dr. E.H. Hamilton of Kilgore to block up some 15,000 acres of leases around and east of Kilgore. Snowden then gave half the leases to Roxanna for a well. Julius Fohs and James H. Gardner, Tulsa geologists, were engaged to make a location. That was the same Fohs Walter knew in Scurry County. The well was too far east by one and a half miles.

However, Gardner says, there was a story, which he could not confirm, that the heavy rigging was unloaded at Kilgore in a driving rainstorm which made the road to the drill site impassable. Jack Shannon of Snowden asked the Roxanna supervisor for permission to drill at a nearby dry location on the edge of Kilgore. The Roxanna man refused. If he had not, and if the story is true, the Woodbine oil would have been discovered 15 years earlier, neither Roxanna nor the East Texans would have succumbed to the depression and Shell might have shoved Standard of Jersey right out of first place in the oil big leagues.

Several other tests were drilled in the vicinity of the field. One of these was east of Longview even before Barney Skipper started hawking oil. Humble, Amerada and Shell were all drilling core holes and some tests atop the Sabine Uplift too far east. Every time any of the big companies rigged up a well east of Longview, Barney would say, "Try just one little old well west of Longview. It might change your luck." But the big company geologists said, "No."

One test which almost threw cold water on the Joiner operation was a Texaco try a few miles south of Henderson that not only was dry but even failed to find the Woodbine above the water level.

All of this history had little to do with 450 acres on the Lathrop lease in Walter and Barney's tract west of Longview. The Joiner well was at least 25 miles to the south. There was no such thing as an oil field that long in this country or anywhere else.

Walter had studied the geology of the East Texas basin. It covered an area from Trinity, Madison and Robertson counties on the south up to the Red River on the north. Its width was from Tarrant County (Fort Worth) on the west right up to the edge of Rusk, Gregg and Upshur counties on the east. But it stopped there. That's why Walter told Farrell he had no geology at all. It's all he could say. There was no structure. He said it might be a stratigraphic trap, a term he picked up from his friend, Leon English.

The cause of interest in the basin had to do with Humble Oil and Refining Company's discovery of production on the southeast flank of Boggy Creek Dome in 1927 and Pure Oil Company's Van field discovery in 1929. The only two reasons for most independent wildcatting in the whole East Texas basin were suspicion and easy access to free leases. That was why Farrell and Moncrief were getting prepared to drill on the Skipper-Lechner tract. The majors were still looking for salt domes and anticlines or any structure seismology could pick up. Seismology could not detect stratigraphic traps—and still can't.

Two days after signing the contract with Farrell and Moncrief, Barney assigned all the leases in the block to Lechner as he had agreed to do. This was done to simplify lease and title matters and was an indication of the confidence the two men had in one another.

Also, some of the people Barney had taken leases from for nothing were asking for them back prior to expiration because lease hounds working the area were offering $1 an acre. These were trick offers and Walter knew it. The leases would be taken subject to title clearance and really obligated the people taking the leases to no payment whatsoever, unless, of course, they could sell the leases for more.

Then they would exercise what amounted to an unfair option since the landowner was at the mercy of the lessee. This was a new wrinkle a group of Oklahoma prospectors had brought into Texas. In time Texas oilmen would learn the trick well.

Part of Walter and Barney's lease was acreage owned by F.K. Lathrop, manager of the Kelly Plow Works in Longview and a highly respected citizen of the community. Although the land he leased to Walter and Barney was several miles west of the town, Lathrop lived in Longview. He was a mature man in his early 60's and had intended to retire to his spread of land. His assistant, J.W. Elliot, looked after the land, which consisted of almost 450 acres. The test well was to be drilled on his lease.

Shortly after the signing of the contract with Farrell and Moncrief, Walter realized that if all titles to the land covered by the 9,300 acres of leases were to be "cured," he would have to have help. Barney had already admitted he knew nothing about title work. It was a job for a specialist. Walter had too much else to do. They had no money with which to hire a title man. Their only hope was to lure one with an offer of acreage. Barney and Walter decided to split their 4,300 acres three ways, giving the title man a third of the leases.

The first thought Walter had was that Jimmy Nowlin, Ruth's brother, who was an outstanding land and title man for Atlantic Refining Company, might be interested. It would mean giving up a fine job in the middle of a rapidly deepening depression and for a third of 4,300 acres in a wildcat area 50 miles from production that had been condemned by almost every expert in every major company in the business.

That was exactly how Nowlin saw the deal. He had always wanted to become a wildcatter, but not under the conditions his brother-in-law had to offer. He said he had better stay on Atlantic's payroll since someone in the family had to have a job. After some persuasion, Walter agreed. Although Atlantic's own East Texas geologist, Bob Whitehead, believed in the prospect, he couldn't blame Jimmy for turning it down.

The Nowlin decision still left Lechner and Skipper without someone to run down titles to their leases. Time was slipping up on them. Finding a man with the qualifications would be difficult with no more incentive than a chance to get rich with acreage in what was generally regarded as an impossible wildcat prospect.

There was another man, however, that Walter could try, his office partner in Dallas, Ray Hubbard. They were both pretty broke. Ray worked for himself and had no salary to hold him. He was a trained title expert who had worked in Oklahoma with the Gypsy Oil Company, a Gulf Oil subsidiary.

Walter and Ray had met on a double date before Walter married Ruth. Ray had a date with Ruth's sister, Marie Nowlin, on that occasion. Later, when Walter came back from the Grimes County leasing expedition that Jimmy Nowlin had lined up for him, he renewed his acquaintance with Ray. Ray then invited Walter to move into his office in the Kirby Building in Dallas. In fact, Ray's father was already in there with him, but Walter could help with the rent.

Ray had been more or less successful in keeping his head above water financially and would require no expense money or other payments. Walter explained the deal to Ray and it appealed to him. He was to get the same one-third interest in the 4,300 acres as Jimmy Nowlin was offered. Walter was happy to get a man of Ray Hubbard's experience, especially to be the inside man, if oil were found, while Walter ran the field operations.

An incident came up about this time in 1930 that would haunt Walter for the rest of his life. Taylor Lee, who had leased some of his land to Walter and Barney, offered to lease up 16,000 acres and turn it over to Walter if he could get someone to drill a well anywhere on the block. It was exactly like the Skipper deal. Lee and his brother were outstanding citizens. He owned a cotton gin, had a large landholding of his own and was successful in the mercantile business among other activities.

Walter was tempted. But he realized that already he was going night and day on the Skipper deal and there was no time for the Lee project. It would be unfair to the Lees as well as to his own partner for him to overextend himself. With regret he turned it down.

It was a most unfortunate, but necessary, decision. The entire 16,000 acres, which started on the west side of Lake Devernia where the Lechner-Skipper leases stopped, turned out to be productive. In his later years Walter realized he had turned down an opportunity that might have made him the biggest individual oilman in the world, Hunt and Getty combined. The profit from those acres probably exceeded a billion dollars.

With land titles completely uncertain, the Lee deal might have been an impossible undertaking for Walter and Ray. Many parcels of land had never had their warranty deeds put on record. Many owners had traded livestock—or even slaves—for land. Often land was won in card games or horse races. Occasionally a man would leave the county and give his land to a neighbor or a black tenant farmer or helper. Much of it had been left to heirs who had never been found. Much had been claimed by squatters. Thousands of acres had been leased about 15 years earlier to the Virginia Oil Company, a promotional group which had since left East Texas and probably was no longer in existence.

It isn't good for a man to look over his shoulder, but now and then, even 50 years after the field came in, Walter thinks about turning down the 16,000 acres of leases Lee offered him, most of which was still productive in 1977.

Skipper, Lechner and Hubbard weren't the only people in trouble west of Longview. Farrell, Moncrief and their associates were bothered by several problems. There were the titles that Walter and Ray would have to clear. Equally important, however, was money.

Neither Farrell nor Moncrief was burdened with an overabundance of cash. That problem was compounded by an almost total absence of credit due to the depression.

They were able to relieve some of their financial troubles by making a deal with the Arkansas Fuel Oil Company for an interest in their acreage in exchange for supplies, services and cash. From time to time they would make other deals with the same company or Tidewater Oil Company or others for acreage out of their block.

They gave Frank Foster a contract to drill the well for $10,500 cash and 500 acres of their leases. But when the time came for the well to get started, Foster had no rig available due to the action in Rusk County. Yet he had an iron-clad contract with Farrell and Moncrief.

To relieve his situation, Foster went to R.L. Foree who owned and operated the Seminole Supply Company of Seminole, Oklahoma, and Electra, Texas. Foster said he would give Seminole half interest in the losses and/or profits that might accrue in contracting the well. Foster and his partner (Jeffries) were to pay the freight on the rig to be shipped out of Seminole and would "look after" the drilling of the well. No money was escrowed. Foster provided a map that outlined the 500 acres near the Lathrop lease.

Seminole (composed of Foree; his vice president, F.A. Turner; and secretary, E.A. Ellison) had made its entrance into the East Texas area a few weeks earlier by selling drill pipe to Ed Bateman of Fort Worth, who was drilling his Lou Della Crim No. 1 well south of Kilgore in Rusk County.

This entire venture, which was to result in the largest oil field in the world for many years, was a "pore boy" operation. Everyone had to approach it with tremendous doubt because all the major oil company geophysicists, geologists and executives had declared it impossible to find oil there. The basis of this belief was the absence of subsurface structure to provide traps for the accumulation of oil and gas.

The deal for the rig with Seminole Supply did not include line pipe or gas and water. Foster made a deal with Don Hugus of Tyler for the loan of enough 2-inch pipe to service the well. For that, Hugus received two 10-acre oil and gas leases.

Once during this time Walter was in Dallas and had car trouble. His funds were extremely low. He offered a service station mechanic a 25-acre tract near the Lathrop lease to do the $30 job. The man told him he had seen too many oil promoters in his time to fall for such a deal. Then Walter made an offer to the car dealer, Bob Bryant, for 100 acres to skip a $25 payment. He also refused. Both turned down land worth millions. Bryant said he had been "had" too many times trading for oil leases and had no confidence in anything connected with the oil business.

Walter, Barney, Hubbard, Farrell, Moncrief, Eddie Showers and the drilling contractors were all scrounging and giving away fortunes for the barest of living and operation necessities.

On September 15 Walter assigned the Lathrop lease to his friend, Johnny Farrell. On September 30 Lathrop, in return for a commitment from Walter to drill the test well on his land, granted Walter one-fourth interest in his one-eighth royalty in all four of the Lathrop tracts. The transaction was recorded in the deed records of Gregg County on page 48 of volume 63 of file No. 1358 when it was filed on December 29, more than three weeks after the well was spudded by Farrell and Moncrief. Mrs. Skipper, Mrs. Lechner, Farrell and a few others received fractions of the quarter of Lathrop's royalty. Farrell gave a small fraction of his Lathrop royalty to Moncrief, who gave it to Ira Rinehart, a friend and publisher of oil reports.

Farrell had verbally granted Walter the right to select the tract to be used for the first well. Walter wanted the privilege for two reasons—as a means of partially compensating Mrs. Skipper for her kindnesses during the leasing program and the period when he was seeking someone to drill a well to fulfill the obligation to the landowners. He also wanted Ruth to have an interest in the first well.

Sometime in October, Farrell and Moncrief went to Longview to inspect the Lathrop lease. Although there was no airport in the little town and Skipper had advised them against coming in a plane, as typical of some oilmen of the day, they decided to do so anyway. They landed on Bivin's Ranch, southeast of Longview. Skipper, upset over what he considered a show-off act, flagged them in.

It is interesting that Lathrop ended up making the location for the well himself, accompanied by Eddie Showers, a partner of Farrell and Moncrief. When stepped off from the east boundary, the location as originally spotted on the map wound up in the middle of a creek. Lathrop then took it upon himself to move it to the east side of the creek. The creek, incidentally, supplied water for the boilers for the drilling operation when it started and became part of the slush pit.

On October 5 the momentous news that the Joiner well had come in was shouted from one end of East Texas to the other. The well opened an oil field in an area where the "geniuses" all said there could be no oil. Walter didn't even have time or sufficient interest in the well so far to the south to go down to see it. Van field was not much farther to the west. It meant little, as he saw it, to the chances for the Lathrop.

Joiner's well came in from the Woodbine sand that had been encountered at a depth of 3,536 feet. That was almost exactly the depth at which the "fake" geologist, old Doc Lloyd, said four years earlier oil would be found on the Daisy Bradford.

It produced 90 barrels of oil "by heads" the first day. That meant it flowed in spurts instead of constantly. The second day it made 87 barrels of oil. It was rated as a 400-barrel-per-day well. It was not a gusher by any means, but neither Dad Joiner nor Doc Lloyd had ever said it would be.

A multitude of farmers, merchants and landowners waited jubilantly for the well to come in. When it did, a carnival atmosphere took over. Nothing like it had ever been seen in that area. Cars, trucks, wagons, horses and buggies jammed the dusty roads around the well. Boys sold soda water and sandwiches. And big company men looked on in disbelief.

When Walter got the flow details on the well, his first thought was that he hoped the Lathrop well would be better.

The Joiner discovery did cause Farrell and Moncrief a bit of trouble. It was because of that well that Foster and Jeffries took a contract to drill on the Ashby lease, almost due east a short distance over the Joiner block boundary. It was being drilled by the Deep Rock Oil Company. Then Foster had to make the deal with Seminole for another rig. Before the Lathrop spudding date, Jeffries and Foster parted company and Seminole had to come in and collaborate on the continued operation of the Lathrop test.

On October 13, before the Lathrop spudding, a man by the name of W.F. Priest, who had received one of Skipper's 750 letters, came in to demand an interest in the block. Priest and Skipper had signed a document in April before Walter joined him. Priest was joined in his complaint by Janice Dotson of Dallas, a secretary who had notarized the affidavit in which Priest swore he had accepted Skipper's offer.

Walter, who had not heard of the April deal before, got Angus Wynne, his attorney, to handle it. Wynne said from the outset the agreement was not binding. This continued to give Skipper, Walter, Farrell and Moncrief trouble until Walter finally settled it some months later by arranging for Skipper to give Priest and Dotson a few acres of leases.

When Skipper and Lechner brought Hubbard into the deal and split the property three ways, Lechner and Hubbard decided to combine their holdings as a joint venture. Skipper took his one-third and went on his own. It was agreed at that time that Walter and Ray would stay to the west of an imaginary line, that Skipper would go east and south, that Tracy Flanagan, leasing for Shell, would go north and that none would invade the others' domains.

During this time, Barney was having an economic crisis. He was completely broke. He didn't even have enough money for groceries and he owed several hundred dollars on gasoline for his car, which he had to have to get around. Once he offered Humble his entire holdings for $3,000 because that was the amount he owed. Humble considered the matter, but reported back in a few days that their geologists had recommended against it.

On December 3 the Lathrop well was finally spudded in, that is, started making hole. The start of operations had been delayed when the United Gas Company in Shreveport refused to connect the well with gas for fuel. Bob Foree of the Seminole Company settled that by a call to the Shreveport headquarters of United.

The well was later designated the Arkansas Fuel Oil Company F.K. Lathrop No. 1-A. It was located on an 850-acre tract which John Farrell assigned to the Arkansas Fuel Oil Company on January 9.

All the principals were there for the spudding—Walter, Barney, Hubbard, Farrell, Moncrief and their families and their partners (Eddie Showers, Bill Noble and R.S. Baker), Foster and Foree as well as representatives of Arkansas Fuel Oil, Tidewater Oil, the town government of Longview and the Gregg County Chamber of Commerce. The Chamber of Commerce had offered a $10,000 bonus for the first well to come in within a certain distance of Longview. The Lathrop location qualified for the reward if it made oil.

Less than two weeks later, Deep Rock's Ashby No. 1 well came in about 6,000 feet west and slightly north of the Joiner well. It made 1,635 barrels of oil in the first 13 hours. It was a good well. When it came in, it was revealed that Foster was not only the drilling contractor but a partner with Deep Rock.

As soon as this well came in, the soon-to-be-fabulous H.L. Hunt made his move. He bought out Dad Joiner, who was hiding in a Dallas hotel almost afraid to show his face. He had oversold the interests in his leases around the well. It was fear of consequences from this that caused him to sell. Hunt was willing to take the risk of a flood of suits over spurious certificates as well as untested titles. It was this move that started him on his way to one of the greatest individual fortunes in American history.

About 12 or 13 miles north of Deep Rock's Ashby No. 1, on the northern boundary of Rusk County, another drama was unfolding. Ed Bateman, a former Houston newspaper advertising man who had gone into the oil brokerage business, had spudded a well on the Lou Della Crim land south of Kilgore in Rusk County. This was in an area where Malcom Crim, the sage of Kilgore, had blocked thousands of acres several times and begged oilmen and promoters to come in and drill. A gypsy fortune teller had told him years earlier he would find oil on his mother's land and he believed her. Since then, he had thought of little other than his small business and the oil that would some day be found. Malcom smoked a corncob pipe and wore a big, flapping felt hat. No one would have taken him for a city slicker.

When Bateman's men went to him and promised to drill a well if he could get 1,500 or 2,000 aces of land, he quickly lined up just under 1,500 acres. The well was drilled on the Lou Della Crim lease on a location selected by Bateman's geologist.

When this well was started, only one major company or any other type of oil company seemed to have the slightest interest in it despite the Joiner and Deep Rock wells. They were considered freaks and certainly a well 13 miles north would be far out of the field, even if there were other wells around what was already known as Joinerville. But Humble had some large lease holdings nearby and assigned scouts to watch progress on the Bateman.

When Bateman's Lou Della Crim Well No. 1 came gushing in some 640 barrels of oil in the first 40 minutes on December 28, 1930, a new boom was on around Kilgore. Some shock waves reached another 13 miles north where the Lathrop well was drilling, but still there was no additional activity between the Joiner and the Bateman wells.

The idea of two fields in the previously condemned mass of hills and valleys and farms and piney woods of East Texas was never considered. No one even gave a thought to one great field, even after the Crim well came in flowing from the same Woodbine strata as the Joiner well, only slightly deeper.

Most geologists and geophysicists and oil executives dismissed the idea of a hook-up between the Joiner and Crim wells. Their lack of vision and stubbornness were going to cost their companies untold millions of dollars and make the sailing much easier for the wildcatters and boomers.

The Crim well wasn't 24 hours old before the first train load of prospectors and camp followers unloaded at the railroad station in Kilgore.

The Bateman well got Walter's attention. He and Barney took an hour off and drove down to Kilgore to see it. His first thought was that he was sorry Barney had not been trying to build Kilgore instead of Longview. Not even Walter, of course, was "silly" enough to think it very likely that the Woodbine blanket of oil would spread all the way to the Lathrop lease.

Barney assured Walter that the Lathrop well would be a producer. He said, "Be patient." They laughed and agreed that Barney was right as they raced Walter's little Ford back to Longview.

While the drilling of the Lathrop was progressing in a highly professional manner, the chaos grew around the site as disputes over titles became more numerous. The operators, Farrell, Moncrief et al, were beginning to get impatient with some aspects of the Lechner-Skipper deal.

For one thing, neither Walter nor Barney had ever become very well acquainted with Moncrief. They had started off on the wrong foot at the signing of the contract when Moncrief and Walter had their insignificant tiff.

Certificates of title are simple statements that the lessors own the land involved. They were easy to obtain and Walter had them for most tracts. When Ray Hubbard came in, he found he had undertaken a most difficult task, which was checking titles for defects. He did a good job and the operators were impressed with the efforts being made to clear titles.

There was, however, a lawsuit and then a cross action and a settlement that didn't help feelings between the parties although Farrell and Lechner never stopped working together and remained friends. There was no enmity between any of the parties, as a matter of fact, simply a coolness between Lechner and Moncrief, two fine men, that seemed to resist any efforts toward reconciliation.

As the drilling of the well progressed, December became an exciting month. The Deep Rock well confirming Joiner's discovery on the 20th and the discovery south of Kilgore in the Bateman well on December 28, to bring in what was then considered another new field, naturally gave hope to the parties interested in the Lathrop well. What these things meant was that the experts had been wrong in their conclusion that no oil could be found in the East Texas basin between Henderson and Tyler or Longview.

Despite their financial troubles which were forcing them to give up valuable acreage out of their block to raise operating funds, Farrell and Moncrief were obviously in a more optimistic mood. Barney had never had the slightest doubt that oil would be found west of Longview.

Walter kept thinking of Dr. Hugh Tucker's prediction about the greatest oil field in the country being found west of Harrison County. Nor could he forget Bob Whitehead's warning that he should never give up because he was "sitting on a keg of dynamite." Both were fine geologists. In fact, Tucker had staked the location for the Yates field discovery, one of the most prolific fields in Texas history.

But times were getting hard. Walter already had difficulty paying his car notes. Now he was having trouble meeting the low rent on his room in the Gregg Hotel. It was a fine room, right at the top of the stairway from the lobby. He had taken Ray Hubbard in as a roommate to help keep down expenses. Both Barney and Walter were busy taking new leases for themselves. Walter, of course, was splitting his leases with Ray.

Barney was getting more ragged as the days went by. As the weather grew colder, he was forced to put cardboard in his shoes to keep them dryer and warmer. Mary's clothes were patched. If it hadn't been for his 900 acres of purple peas and the low cost of chickens, food would have been a real problem. Barney's charge account with the service station for his gasoline was getting to be a real burden on the patient dealer. Occasionally when Walter would go to the Skipper shack for one of Mrs. Skipper's delicious breakfasts, they would all kid each other about their patched clothes. Walter hadn't bought a new suit or a shirt in more than a year.

Occasionally Jimmy Nowlin would offer to loan Walter a few dollars, but he refused, knowing that Jimmy was newly married and needed his money.

Nowlin was a typical oilman of his day. He knew everything there was to know in the field. He could direct production activities or land acquisition or title work. He had once been a baseball player. In his younger days he had been quite a lady's man because he was handsome and clever and had a fine personality. His name was Wayne Ivis Nowlin, but one of his early girl friends didn't like the names so she called him Jimmy. It stuck for the rest of his life.

He would occasionally pay Walter a visit and offer whatever help he could. He turned down Walter's offer to make him a millionaire, but he had no regrets. He was seriously worried about Walter's welfare, however, when he saw him in tattered clothing on a visit to Longview shortly before the Lathrop well came in.

The excitement around the Lathrop well started in the first two weeks of January. One day Walter was handed a small piece of a core from the well and he wrapped it in a handkerchief without anyone's observing. It was said Farrell directed a crew member to make the delivery, since it was almost impossible for Walter or Barney to get on the derrick floor. They were certainly not welcome by either Moncrief or the drilling crew.

When Walter took the brown-stained handkerchief out of his pocket, it had a strong oil odor. Then he knew he had finally found his pot of gold at the end of the rainbow. He went back to Longview. The first person he saw was his sister, Lucille Thrasher. She was sitting in a car in front of the post office. He took the handkerchief out of his pocket and told her to smell it.

"That, sister, is oil sand," he said.

Lucille lived in Longview at the time. Frequently Walter spent the night at the Thrasher home to avoid landowners who wanted their leases back as the drilling progressed.

Then he told Skipper about the sample, but Barney was skeptical. He thought someone was fooling Walter. On January 13 the news leaked out that a rich oil sand had been cored in the Lathrop well. The *Longview News* carried a banner headline proclaiming, "Oil Sand Found In Lathrop Well." Even then Barney was persuaded by Frank Foster and Eddie Showers that the report was not accurate.

In fact, he was led to believe, by indirection, that the report was simply a ruse to unload acreage to recover the lease and drilling investment. That wasn't unusual, but Barney, a man of high integrity and deep religious faith, was not one to engage in such nefarious practices even when he was offered $3,000 for the acreage he controlled. He was inclined to believe the story in the newspaper was wrong because only a week or ten days earlier Humble Oil and Refining Company had turned down his offer to sell out.

Walter did his best to boost Barney's spirits. He knew there was oil under the Lathrop tract because of the sample cutting from the hole and he knew that probably every acre of land from just west of Longview for miles toward Gladewater and Tyler was also good. In fact, Walter was almost certain every acre of land between the Joiner and the Lathrop and for miles to the north would be good. He now had total confidence in Hugh Tucker's pronouncement about the great oil field.

For the first time in years, Barney had begun to lose his faith in Gregg County oil. But Walter spent as much time as possible dispelling his gloom. He got Barney to go with him and watch the well from a distance. He pointed out indications that were clear to a trained oilman that no one was preparing to abandon the Lathrop No. 1.

That helped. Had it not, he might have sold his leases for $3,000 on the 25th to two men. They told him outright the well was dry. One of the men asked Mary if she wouldn't like a new Easter bonnet and some clothes. Her answer was simply that she didn't know them and couldn't understand why they would give her husband $3,000 for a block of oil leases that were worthless. She also asked something Walter had suggested she might ask. That is, why hadn't they offered Lechner and Hubbard anything for their leases?

"They know," Walter had said, "that no oilman would be trapped into such a deal for worthless land. They also know Barney knows very little about the business."

That gave Barney a boost. He finally told the men he had waited for a well to be drilled for 25 years and he could wait a few more days.

Walter was being bothered by so many landowners trying to get their leases back that he went to Marshall without saying anything to anyone. He was convinced his days of oil-field poverty were over and suspected that the well was on the verge of coming in as a producer.

It was in the early hours of January 26 that someone came beating on Barney Skipper's door, shouting that the well had blown in and was flowing oil.

Barney and Mary jumped out of bed. Barney was so excited he put his trousers on backwards. Then he knelt down beside the bed and thanked God. He went immediately to the well and there he saw the two men who had tried to buy his leases. He started toward them with his 6-foot, 200-pound frame. They ran. Barney said later all he wanted to do was tell them they had been wrong.

Within a few hours, hundreds had gathered at the well site. As the day dawned, the crowds grew. By afternoon, more than 5,000 people cheered as the oil flowed. The crew fought to control the flow in order to prevent an accident by a careless smoker or to prevent someone's being drowned in a slush pit full of oil.

The well was gauged at 18,000 barrels of oil per day. It was a better well than the Joiner, the Deep Rock or the Bateman wells. The estimate was based on a controlled flow of 133 barrels of oil in one hour through 2½-inch tubing. The *Longview News* headline proclaimed the well a gusher. The little town, soon to be a city, went wild.

Walter heard the official news on the radio in his hotel room in Marshall. He knew that now was the time to buy and sell. He had some leases he would have to sell because he needed expense money as well as money to pay back debts. He also needed money to purchase some new leases, some even further from the well. He had no intention, of course, of selling his leases near the Lathrop. In fact, he bought several leases nearby for $5 and $10 an acre.

Today for the first time in some years Walter felt he was financially secure, although he still had no money. Shortly after the notice of the core test, which was never confirmed by the operators, lease prices had taken quite a jump. With the well coming in, it was said that some leases would jump from zero to thousands of dollars an acre. That didn't happen for a while.

One of the first things Walter had to do, since all the 9,300 acres had been assigned to him by Skipper two days after the Farrell-Moncrief contract was signed, was to transfer Barney's one-third of the 4,300 acres back to his name. The 5,000 acres had not been finally assigned to the operators when the well blew in, but were in a few days.

Skipper had other leases, those he had made since the division of the acreage, and most of them were in the high lease range. He and Mary were in a position not only to pay off their debts but to start looking for a better house and some new clothes.

A few days after the well came in, the Chamber of Commerce awarded the bonus to Farrell and Moncrief. They divided it among the members of the drilling crew who had done a magnificent job of drilling and completing the well.

Walter waited a few days to return to Longview. When he did, he and Ray continued to clear up titles for Farrell and Moncrief as well as those for the Lechner and Skipper tracts. There was the matter of transferring the 5,000 acres to the operators. He did, however, place a lis pendens (notice of pending lawsuit) on the property before the transfer.

This became highly significant some weeks later when Farrell and Moncrief decided to sell what they had left of the 5,000 acres to the Yount-Lee Oil Company of Beaumont. That company had become one of the most remarkably successful independent organizations in the country. It had brought in the well that started the second boom in the famous Spindletop field where it had leases on most of the acreage. The profits had been nothing short of fabulous. The company had found a dozen or so other good oil fields, mostly along the Texas-Louisiana Gulf Coast.

Therefore, it had the cash to provide Farrell and his partners with about a million and a half dollars as a down payment on the acreage. Yount also agreed to pay another $2 million out of one-fourth of the oil produced from the property.

But Walter's lis pendens delayed the sale. A.D. Moore, a capable young attorney for Yount-Lee, refused to close until Walter had waived his claim. That was accomplished in a rather daring race with time by Eddie Showers from Beaumont to Longview to file all necessary papers in time.

For agreeing to sign the papers, Walter and his associates, Ray Hubbard and F.L. Luckel, were assigned one-fifth of the $2 million oil payment from Yount-Lee to Farrell and Moncrief et al until a sum of $150,000 had been paid. It was an exciting and rewarding experience for Walter, who saw the end of the lawsuits between himself and his old friend, Johnny Farrell, and his associates with a financial victory. Walter bought Ruth a new home in Dallas.

For property for which they had agreed to pay Lechner and Skipper $2,500 and a well a few months earlier, Farrell, Moncrief et al received a total consideration of $3,270,000. The deal was closed on March 11, 1931, little more than two months after the Lathrop was spudded. That included the $2 million oil payment.

While the division of the cash payment was not revealed, the breakdown of the proration of the oil payment showed that Farrell was to receive 50 percent and that Moncrief and his associates were to receive the other half, including 18¾ percent for

Four-year-old Walter Lechner and his favorite dog Jack.

Walter William Lechner at six months of age.

About 1910, Walter and his co-workers in an automotive dealership in Dallas sold the Thomas Flyer among other early models.

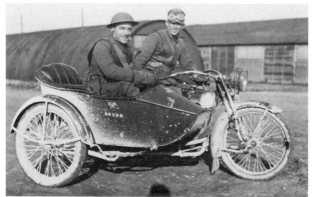

Lechner and driver "Bill" Williams in side car in France in 1918.

Five-year-old Rembert Andrew Lechner, Walter's only son, enjoys his perch in his daddy's Ford.

Mrs. Ruth Lechner drives the stake at the J.J. Moore #1, south of Ira in Scurry County, on January 2, 1923. Mr. Moore is holding his hat. The well was spudded February 5, 1923.

Digging water well for J.J. Moore #1.

Part of W.W. Lechner's casing crew. E.I. "Tommy" Thompson, E.E. "Buddy" Fogelson, and W.W. Lechner running pipe on the J.J. Moore #1 well.

The completed J.J. Moore well, first oil producer in Scurry County.

A church yard full of oilwells—14 in this view—in downtown Kilgore, Gregg County, East Texas Field. *(Photo: Longview Chamber of Commerce.)*

J.J. Moore and his son, Pat (on right), dig the cellar for the rig.

Barren waste and T-model Fords around J.J. Moore #1.

E.I. (Tommy) Thompson and Buddy Fogelson, associates of Lechner, inspect a lease in McCulloch County.

Sign on door at Loutex office in Snyder, Scurry County.

Wagons jam the highway leading to Burkburnett in 1918 following the discovery of oil at a well called Fowler's Folly.

The greatest of East Texas oil well fires claimed nine lives when the Sinclair No. 1 well on the Cole farm burst into flames in 1931. *(Photo: Jack Nolan.)*

The J.J. Moore #1 responds to a nitrogen shot.

Ranger, Texas, about 1919.

Cars blocked the road leading to the Daisy Bradford #3 when Ed Laster was trying to bring in the well. *(Photo: Texas Mid-Continent Oil and Gas Association.)*

Promoters, oilmen and all sorts of camp followers mob the streets of Desdemona during the boom of 1918.

Barney Skipper, driller Andy Anderson and F.K. Lathrop in front of the Lathrop 1, in Gregg County.

Doc Lloyd, Daisy Bradford and Dad Joiner in front of the soda water stand Miss Daisy's nephew built to quench the thirsts of the thousands who witnessed the birth of the "Black Giant" in 1930.

Driller Ed C. Laster, without whom there would have been no Joiner well.

Of the 24 wells drilled in the 1930's on the "world's richest half block" along Kilgore's Commerce Street, only one stands as a commemorative marker. The rest of the half block is a parking lot. *(Photo: Kilgore News Herald.)*

January 26, 1931—Historic Lathrop well (Gregg County, west of Longview) comes in to prove East Texas field is a single giant reservoir instead of a series of individual fields. *(Photo: Longview Daily News—Longview Morning Journal.)*

120

WESTERN KB (23) UNION

The filing time shown in the date line on telegrams and day letters is STANDARD TIME at point of origin. Time of receipt is STANDARD TIME at point of destination.

Received at W. U. Bldg., Cor. Main & Pearl St., Dallas, Texas Always Open

ISC49 64 DL=SC AUSTIN TEX 26 410P

D57 NOV 26 PM 4 27

HON W W LECHNER=

        DAL=

I SHOULD LIKE TO APPOINT YOU TOGETHER WITH HONORABLE LLANO
MATIAS OF LAREDO AS MY OFFICIAL REPRESENTATIVES AT THE
INAUGURATION OF GOVERNOR ELECT RODRIGUEZ TRIANA AT SALTILLO,
COAHUILA, MEXICO, ON THE MORNING OF DECEMBER 1. PLEASE
ADVISE ME YOU WILL ACCEPT. SENATOR ALBERTO SALINAS CARRANZA
WILL BE AT HOTEL HAMILTON LAREDO TEXAS SUNDAY NOVEMBER
TWENTY EIGHTH IN EVENING TO MEET YOU AND LLANO MATIAS=

        JAMES V ALLRED GOVERNOR OF TEXAS.

Allred telegram requesting Lechner to represent him at Saltillo.

Gov. Rodrigo Triana (center of group in big hat), Lechner to his left, Senator Alberto Salinas Caranza in front of Lechner, Col. Matias de Llano at extreme right.

Lechner entertains 169th Aero-Squadron veterans at Eagle Mountain Lake lodge in the 40's.

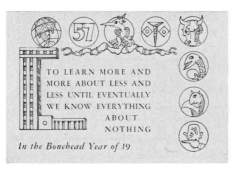

Walter Lechner hams it up in Bonehead Club of Dallas in the late 40's. *(Photo: Neal Lyons.)*

The Texas Independent Producers and Royalty Owners Association (TIPRO) was formed in 1946 to oppose a proposed Anglo-American oil treaty and to promote termination of wartime price controls on crude oil. The first board of directors included (standing) Walter A. Henshaw (San Antonio), Bryan W. Payne (Tyler—later a president), A.E. Hermann (Amarillo—later a president), Walter W. Lechner (Dallas), John W. Naylor (Ft. Worth), and Edward G. Kadane (Wichita Falls); (seated) E.I. (Tommy) Thompson (Dallas), H.J. Porter (Houston—founding president), and Glenn H. McCarthy (Houston).

Gov. Beauford Jester (seated) signing a bill about 1946. Watching him are Howard Dodgen (Austin), then executive director of the Texas Game, Fish and Oyster Commission; Clarence Jones; and Texoma Commission members, Rep. Cliff Gardner (Gainesville), Walter Lechner (Dallas), and Joe E. Cooper, chairman (Dallas).

Digging first spadeful of dirt in June, 1954, for the canal to open Rollover Pass are Terry Scarbrough (Kennedy, Texas), Henry Coffield (Marfa), J.W. Elliot (Mexia), members of the Texas Game and Fish Commission; H.D. Dodgen (Austin), executive secretary; Herbert Cole, president of Beaumont Sportsman's Club; Henry Le Blanc (Port Arthur), commission member; and W.W. Lechner (with shovel), chairman. The pass was officially opened April 23, 1955.

(Photo: Estil Linnens.)

Bill Decker

Picture autographed to W.W. Lechner by his close friend, the late Speaker Sam Rayburn.

Walter and Ruth Lechner cut their 50th wedding anniversary cake September 1, 1970.

Gov. James V Allred

Allan Shivers

Lyndon B. Johnson
(Photo: Arnold Newman.)

Photo of Walter
Lechner taken for
Dallas Times Herald
November 1, 1970.

W.S. Noble, 12½ percent each for Moncrief and E.A. Showers and 6¼ percent for R.S. Baker.

Forty-two tracts comprising some 3,500 of the original 5,000 acres were involved. The remaining 1,500 acres included the 500 acres to Foster (who gave half of it to Seminole for the drilling rig and equipment) and the acreage given to Arkansas Fuel Oil Company (now Cities Service), Tidewater Oil Company (now Getty) and others for expenses in bringing about the second 13-mile extension to the East Texas field.

Therefore, in less than six months following the contract between Farrell and Lechner, all involved—including those who put up the money, the land, the supplies and the services—had become wealthy because Lechner had never lost faith in Hugh Tucker's vision and Skipper had never turned away from his own belief. Now each had found his own reward for faith and persistence. It would be a while before Skipper and Lechner cashed in. Ahead were some nervous days and days of doubt that would require courage and additional faith to ride out.

By the middle of June, 1932, a dozen significant wells had been drilled to outline the great field's limits of 44 miles long and 14 miles wide. After the first four, which were the Joiner, the Deep Rock, the Bateman and Farrell and Moncrief's Lathrop well, the others in order were:

No. 5—March 13, 1931, East Texas Refining Company No. 1 C. Fisher, 4 miles south of the Lathrop.

No. 6—March 30, 1931, the Guy Lewis No. 1 Cash, about 5 miles west of the Joiner.

No. 7—April 1, 1931, Harter and Gaskey's Anders No. 1, about 6 miles north of the Bateman.

No. 8—April 9, 1931, Selby Oil and Gas Company, No. 1 Snevely Heirs, 8 miles west and slightly south of the Lathrop.

No. 9—May 2, 1931, Mudge Oil Company's No. 1 Richardson, about 5 miles northwest of the Lathrop. It was the first well in Upshur County.

No. 10—May 16, 1931, F. and M. Drilling Company's Blackstone No. 1, about 4 miles southwest of the Lathrop.

No. 11—May 3, 1932, Flanagan Production Company's Flanagan Fee No. 1, about 6 miles north of the Lathrop and in Upshur County.

No. 12—May 21, 1932, also in Upshur County and the most northern of all early test wells, the Maddox Development Company's No. 1 Starr, 2 miles north of the Flanagan well.

The wells were all included in the East Texas Engineering Association's report classifying them as discovery and early extention wells. As this report was being made, dozens of other test wells were either started, staked or brought in. The number of new locations was mounting by the day.

Looking back, many oilmen agreed it should have been named the Joiner field. It was really never named. It was simply called the East Texas field because it covered a region. But the name "East Texas" implied an area from Port Arthur to Houston on the south and from Texarkana to Dallas on the north. The field was only a small fraction of that vast area.

It was of more than passing interest to Walter that of those dozen wells that defined the reasonable limits of the giant field, not a single one of them had been drilled by a major oil company. Humble Oil and Refining Company had come in to buy out the Bateman interests around the Crim well for $1,500,000 cash and a $600,000 oil payment which turned out to be dirt cheap as had the Farrell and Moncrief sale to Yount-Lee.

Many major companies complained of a money shortage because of the depression, but a shortage of money hadn't kept the little, broke independents out. Shell came in rather sheepishly, assembled a fairly solid lease block west of the Lechner-Skipper block and then checkerboarded it out to Yount-Lee, which was a company with probably as much ready cash as any company in the country, independent or major. It was a foolish sale, but it was an old Shell tactic of getting back its investment in leases before starting to operate in a wildcat area.

Around the Lathrop well, Amerada, Simms, Tidewater, Stanolind and Sun started operations. Atlantic waited until more than 500 wells had been drilled and then paid tremendous prices for leases to get a foothold in the great field. A few months earlier, by merely following Bob Whitehead's advice, that company could have had the Lechner and Skipper acreage plus untold other thousands of acres in the fairway. Atlantic, with the only geologist who believed in the prospect among all the majors, had simply turned Whitehead down and threatened to fire him if he as much as brought the matter up again. Walter, who often "drank out of the same fruit jar" with Whitehead, was delighted at the turn of events.

Atlantic got started after Whitehead wrote a second letter to Philadelphia, this time stating that it seemed the Dallas office was being run by idiots. That letter got action. Philadelphia sent in 30 landmen and eventually became the field's fourth largest producer.

As the days passed after the Lathrop roared in, space between wells in the East Texas oil map was steadily filled in. The 12 significant outpost tests had proved beyond the slightest doubt that all the wells were from the same Woodbine sand and mostly at within 50 to 100 feet of each other in depth. All the oil being produced was high quality, almost Pennsylvania-grade crude. Five wells had been completed in 1930 and produced a total of 27,135 barrels of oil. In 1931 a total of 3,607 additional wells were completed and produced 109,049,478 barrels of oil. That number almost tripled in 1932 when the first dry holes showed up, one of them Walter's.

Eventually there were 25,976 wells in the field at one time on 130,000 producing acres. Almost 29,000 wells were actually drilled in the field and only 555 of them were dry, probably the highest statistical record of producers to dry holes of any major field in history.

One of the early problems in the field was the lack of a supply and service center. Almost everything had to come out of Shreveport, 60 miles away in Louisiana. Eventually supplies and services were available in several areas of East Texas, one of the first being Willow Springs, west of Longview on the Texas and Pacific Railroad on the road to Gladewater. Later it became Greggton and, eventually, part of Longview.

For about two years before the great East Texas discovery, a dozen of the 100 largest fields in American history were discovered. This rush was due to the passage of the 27.5 percent depletion provision of the federal tax laws. Congress had enacted the law in 1926 in the public interest to prevent the nation from running out of oil.

During an oil shortage crisis after World War I, it was feared that the nation had only a few years' supply of oil, just at a time when the country was beginning to build great highways and

industrialization was in full swing. It was actually believed that
this country would soon become dependent on foreign sources for
its oil supplies. Even seasoned oilmen were then warning their
own sons to get into another business because "all the oil in this
country has been found."

Now, five years later, the nation was faced with an oversupply
of oil, thanks to the depletion law which its sponsors in the
Senate accurately predicted would provide an abundant supply of
oil at low consumer prices.

As early as January 29, 1930, when there was still no hope for
the Dad Joiner well, the Independent Petroleum Association of
America had urged Congress to pass an import tariff. It was the
independent oilmen's contention that the oil surplus was caused
by foreign imports. But since then, new giant oil fields had come
in including Calliou Island and Rodessa in Louisiana. Furthermore, this nation was being flooded by an additional torrent of
oil from Mexico by the El Aguila discovery of the large Poza
Rico field. El Aguila (The Eagle) was a subsidiary of Shell.

Now it was obvious to Walter that, before long, there was going to be overproduction in the East Texas field and that it would
be massive. He and Ray talked the matter over and decided not
to move too fast in developing their acreage, but to keep on the
lookout for more leases to buy with money they had received
from earlier sales of tracts at what they thought were exorbitant
prices. They bought two or three excellent leases for $1 an acre,
including the prolific Dobie, the Hartley and the McGrede. It
cost about $30,000 to equip, service and drill a well in the field.

The Texas Railroad Commission, seeing chaos in the making
in the giant field, issued its first East Texas proration order on
April 4 to be effective May 1, 1931, for 1,000 barrels per well per
day. The order was changed upward twice before May 1.

The Commission was in its infancy in the field of proration. It
had issued its first statewide order on August 14, 1930, with the
effective date of August 27 for 750,000 barrels of oil daily, about
50,000 barrels daily less than had been produced the preceding
year. The limit was based on market demand. By April there
were 235 wells in the field producing about 280,000 barrels of oil
daily. The cutback was relatively slight from the average 1,200
barrels per well. It was the highest per well allowable in the
field's all-time history.

The commission's order was disregarded by most operators. They doubted that the commission had legal authority. Anyway, they were producing so much oil, they felt they could stand whatever fine might be imposed if it were a legal order.

As a result of this lack of discipline, by July 1 the "going price" of oil was 15 cents per barrel and most operators were getting a dime or less. Yount-Lee, rather than pay landowners ridiculously low royalties, simply shut in its wells. Many thought this would result in the company's oil being drained. It would not, Frank Yount said, because migration of the oil was from west to east and most of his wells were far enough east to eliminate such worry.

About two weeks after the Lathrop well came in, Walter and Ray were fortunate enough to get started on their first well on the Tennery lease. It was about three-quarters of a mile northeast of the Lathrop.

Just when they were worried about getting a rig, due to the growing demand, Jim Rush came to their small office just outside Longview on the Marshall highway and said he would like to drill two wells free on the Dobie lease. He was dressed up more like a male fashion model than a drilling contractor and was driving a long Lincoln Continental. Jim's brothers were Joe and Ralph Rush, all in the same drilling venture.

What they wanted was to get their hands on some leases. When he looked at the map and found eight more leases in the block, Jim agreed to drill 18 free wells in all. In view of the high price of drilling and the need for some action as soon as possible, Walter and Ray told him it was a deal. He calculated it would cost Ray and him about $600,000 to drill 18 wells.

The Rush brothers didn't have any money, just rigs. So Jim went to W.L. Todd, who had recently resigned from Simms Oil Company, to get the money. Todd said he would help. He didn't have any money, either, but he did have a good reputation as an excellent oilman known for his ability, honesty and decency.

So he went to Lufkin to see his friends at the Lufkin Foundry and Machine Works. They put up $50,000, which bought them 25 percent of the deal. They simply split with Todd, who got half the Rush deal for getting the money. Therefore, the Rush company became the Columbia Oil and Gas Company of Dallas. Joe and Jim had 25 percent each. Ralph apparently decided to stay out.

Late in February, the Tennery well came in as an excellent producer. Lechner and Hubbard were on their way. Walter was acting as field superintendent and Ray was taking care of the business in the office. They did not have a partnership. Walter had learned somewhere along the way to stay out of partnerships and thereby avoid entangling alliances, so to speak. Theirs was a joint venture.

By the time Jim and Joe had drilled up about half the wells, they found themselves in debt to the First National Bank in Dallas for $325,000. Oil was a dime a barrel. The bank said they had to get the loan cut down. But Columbia couldn't cut it down. The only way they could cut it down was to sell something. They started looking for a half-interest buyer and finally found one in Stanolind Oil and Gas Company, a producing subsidiary of Standard of Indiana. Stanolind made an acceptable offer of about $900,000 of which $600,000 was in cash and $300,000 was in oil payment.

Stanolind offered either to go ahead on the Rush deal or split up the properties. Lechner and Hubbard then had nine completed wells. They decided to split. They took the north half of each lease and Stanolind took the south half.

Jim Rush came back into the picture again and offered to drill the rest of their acreage for $1 a foot. Jim now had money and he still had his rigs. Oil was still selling for almost nothing so Lechner and Hubbard accepted but said they didn't have any money. In fact, drilling had stopped all over the giant field. No one could afford to drill wells on dime oil.

Rush said he didn't need any money. He had plenty. He wanted his rigs working so he could hold onto his crews until the price of oil came back. It was agreed. Ray and Walter set aside for Rush three-eighths of the money from the oil produced until he was paid off.

They then went to Oil Well Supply, National Supply, Continental Supply and Frick Reed companies to get field pipe, storage tanks and other necessary supplies and equipment. Oil Well and National wanted no part of the deal, but Continental and Frick Reed said they would go with Walter and Ray.

They agreed to advance credit with another three-eighths of the income from oil as collateral. That left Walter and Ray with an eighth each until the drilling, supplies and equipment were paid off.

They then engaged Tom Pollard, a lawyer, to obtain as many exceptions to Rule 37 of the spacing law for the East Texas field as he could. They sent him after a location on every spot where landowners demanded wells to prevent drainage by close offset wells, many by major companies.

Rule 37 was the spacing rule applying to a particular field. An exception was a right granted an operator or a landowner to drill on less than the field spacing pattern, for instance, five acres in the giant field.

Getting such exceptions was relatively easy. The Railroad Commission, under the law, had little authority to turn down applications. Lawyers would charge $200 if the exception could be obtained without going to court and $500 if a hearing were necessary.

If Tom had to go to court, it was simple. He would walk up to the judge's bench, even during a trial, and the judge would simply say, "Write up your instrument and I will sign it." That's all there was to it.

When the judge ran for reelection, however, Tom would give a big party for him and contribute to his campaign fund in appreciation. He would also see to it that the judge met voters he didn't know.

Once, on the Dollahite lease, Rush drilled a dry hole, the first ever drilled in the field. The Woodbine was as hard and solid as marble. Jim pulled his pipe, plugged the hole and said, "We just hit a tight place." He moved the rig 75 feet north and drilled a flowing well.

Walter sold a few leases on the outer fringes of the field for from $5 to $25 an acre before things turned for the better. He farmed out some leases to an operator named Turnbow.

One practice going on during the East Texas boom annoyed Walter. That was the way slick oilmen would try to beat landowners out of their royalty. He was especially irritated when they tried to do this to the Negroes and the very poor whites who had no experience in business of any kind, especially the intricate oil business. He also resented and did something about a handful of Oklahoma operators who would hold leases an unduly long time on the pretense of examining titles. He engaged Joe Orr, another lawyer, to fight these abuses. Orr, the clever master in the courtroom, never lost.

In 1932 Walter and Ray formed the Humack Oil Company, a joint venture. They also formed the W.W. Lechner, Inc., and G.E. Hubbard and Son, Inc., two corporations which jointly owned Humack. G.E. Hubbard was Ray's father, whose only interest in the business was through Ray. Humack was a combination of the names Hubbard and McGrede, which was one of the best leases they had.

Walter was never a man for much self-advertising and didn't even want his name to become a part of his own joint venture. He had an obsession for anonymity. Later in life when he was campaigning for political friends or helping in the war effort or serving on state commissions, he relented and permitted the use of his name in the press on occasion.

W.M. Harley wanted to sell Walter his 160-acre farm for about $6 an acre in fee. Walter asked if he would take $5 and Mr. Harley readily agreed.

"Mr. Lechner," he said, "I've been living out here all my life eating black-eyed peas and sow belly and I'm tired."

Walter looked him in the eye and said, "Mr. Harley, I wouldn't buy your land if I could. Hang on to it for a little while longer. You can always get $6 an acre. In fact, I would pay you $6 an acre for a lease now, but I have no cash. If oil is discovered, you can then live in peace and comfort for the rest of your life."

Harley took Walter's advice. Later he was offered a big price for his lease but sold it to Walter for $6 an acre. He prospered and lived the rest of his life in ease. He never missed an opportunity to thank Walter for not buying his land outright.

For weeks after the Lathrop came in, things were hectic. Overproduction was completely out of hand. The Texas Railroad Commission, saddled with oil and gas conservation responsibilities 20 years earlier, was unprepared for East Texas.

From June 7 to August 18, 1931, Yount-Lee was shut down on the block because of insufficient storage for the rampaging flow of oil. On August 17 Governor Ross Sterling declared martial law and sent troops in. All production was closed in. Lawlessness was widespread. By September 5, when the field was reopened, the price of crude oil was up to about 50 cents. Commission allowables were being ignored. The excess over allowables was actually being stolen from the landowners and the state. There was no royalty plan for the so-called "hot oil" and no state, county or municipal taxes were paid on it. Hot oil activities started when the closed-in field sent prices upward.

The field was filled with operators who knew little about the industry. Many were fly-by-night outlaws who would stop at nothing, including violence. They would steal oil by tapping the gathering lines of legitimate producers and then sell out to the ever-growing number of tea kettle refineries. Martial law was in effect until December 11, 1931, when it was declared unnecessary by the U.S. Supreme Court.

The Railroad Commission then took over enforcement and shut the field in for 38 days starting on December 11, 1931. The commission shut the field in three times for shorter periods after that. It was forced to fight in the courts for the right to enforce its rules.

In the meantime Governor Sterling had engaged Ernest O. Thompson, mayor of Amarillo and a distinguished World War I soldier, to fill the sudden vacancy left by former Governor Pat Neff on the commission. Neff had accepted the presidency of Baylor University. Thompson was also a lawyer. The other two commissioners, both much older men, asked him to take over in East Texas. Before long Thompson had done the impossible. Proration was working and the price of oil had returned to normal, around $1 a barrel.

That was in early 1933. Soon all of Walter's 42 wells were flowing and he and Ray were making money. They paid off Jim Rush, Continental Supply and Frick Reed. The money continued to flow in. They were on their way to the wealth most small, impecunious oilmen always dream about.

That was when they decided to take a well-deserved rest. For their manager, they hired Bill Harrison, who was recommended by Todd and Rush. He was a native of Arkansas, a graduate of the University of Georgia and an accountant. He had once worked for Todd as a field superintendent in the Henrietta oil field, when Todd was production manager for Simms Oil.

As soon as Harrison was established in the office on the Marshall highway, Walter and Ray cleaned off their desks and said they were going to take a long trip together with their wives. They didn't know when they were coming back.

Chapter 8

In their earlier, struggling years Walter and Ruth Lechner quite literally did not have time to smell the flowers. And now, in the mid-thirties, they were determined to do so, in a way that would lead to much of what Dallas enjoys today in garden and streetside loveliness.

They began with the first planting in the city of a flowering shrub that, in parts of Texas, was considered as exotic as the African orchid. Today the azalea blooms in a beautiful trail the width of Dallas.

In 1934 Walter and Ruth bought a new, large home at 6921 Lakewood Boulevard and spent the next twelve years there, active and full and graceful years. Shortly after moving there, Ruth recommended that they buy the adjoining vacant lot in order to beautify it.

She knew what she wanted. In Tyler they had seen the colorful azaleas grow. They were disappointed, but not convinced, when landscapers insisted they could not be grown successfully in Dallas.

How to landscape the property was still very much on her mind when the Lechners left, in early 1935, on a transcontinental trip that would fulfill her wishes almost by accident.

By 1935 the Lechners were free to travel, to explore new places, to broaden their horizons. They had worked relentlessly for three years without a rest. Now things were in order. The oil was sold. Money was coming in, for the first time more money than they actually needed to live on. They owned a home and a new Lincoln and they were tired of competing.

Most of all, Walter finally had found a man capable of running his business. He had confidence in Bill Harrison, an honest and enterprising fellow who was to become his strong right hand. He left the company in Harrison's charge and Walter and Ruth took off on the vacation they had so richly earned.

They were joined by Ray and Janet Hubbard, traveling by train to California. Walter had shipped his Lincoln ahead to Los Angeles, where the Lincoln Company furnished them a driver.

Driving along the West Coast toward Portland, Oregon, they noticed frequent signs and billboards advertising the wonders of a place called Lambert Gardens. The signs were festooned with displays of camellias and azaleas in brilliant colors. "When we get to Portland," Ruth said, "the first thing I want to see is Lambert Gardens."

They arrived there to be greeted by the owner, A.B. Lambert, Sr., who showed them proudly through his beautifully appointed grounds. He listened attentively when Ruth sighed, "I want to grow these at home, but I can't get anyone to plant them. Azaleas grow in Tyler in East Texas, but they say they won't grow in Dallas in that black mud where we live."

Lambert nodded. "The soil is different," he said, "but you can make soil. I have a brother in the landscaping business in Shreveport, Joe Lambert, Sr. Leave your address with me and I'll have him get in touch with you."

When they returned to Dallas a month later, they found three telegrams and two letters waiting for them from Shreveport. Ruth called and a few days later Joe Lambert sat in their living room, talking shrubs and flowers. He won them over when he assured Ruth, "Honey, anybody that wants azaleas as bad as you do, they'll grow for them if you prepare the soil."

Joe agreed that the black mud of Dallas would not support azaleas. He suggested digging beds two feet deep, a relatively rare procedure then in home gardening. Impressed, Ruth hired the Lamberts to landscape their home and the adjoining lot. Four Lambert brothers—Edwin, Henry, Gordon and Joe, Jr.—worked under their father. At one time or another, all except Gordon, who stayed close to the Shreveport office, worked on the Lechner project.

From the beginning the garden was Ruth Lechner's project and it became her triumph. Within three years, police were needed in the streets outside to direct the traffic that piled up as motorists paused to admire the magnificent blooms. One Sunday Walter counted the crowd. Over 5,000 people went through their yard.

Letters came in from all over the country, from strangers they had never seen and never expected to meet, thanking the Lechners for sharing the beauty of their garden. Soon other Dallas families began to plant azaleas. The reputation of the Lamberts spread.

The flowers, so natural and uncomplicated, really had touched many hearts. The azalea became a kind of good omen, benefiting all who were involved in the project.

Seeing their beauty so much in evidence today, taken almost for granted, one cannot easily appreciate what an innovation it was 40 years ago.

The first step was to send in a survey team to measure the property and lay it out in all directions. The job had an impact that no one at the time could have imagined.

It was the catalyst for a beautification program that reached across all Dallas, forever changing it from a city of cedars and junipers and drab, colorless northern growth.

The Lamberts gradually were captivated by the city's charm and spirit and they opened an office there that was to lend ideas and direction to the new face of Dallas.

Henry Lambert was assigned to supervise the Lechner gardens, his first major undertaking for the family. His first impression was that the mud was like chewing gum and the trees seemed stunted. The wind was brutal, especially on White Rock Lake. And there was snow. Henry was just a kid. He did not like Dallas at all, but the second year, he tolerated it. The third year he caught himself bragging on it and thereafter he vowed never to live any other place.

It must have been a remarkable sight when Henry Lambert drove through the streets of Dallas at the wheel of a pickup truck loaded with 129 beautiful azalea plants.

They were probably the first azaleas in Dallas, certainly the first mass planting of them. They attracted great attention. But for a long time people still did not believe they would grow.

The Lamberts worked the beds with peat moss and sandy loam, creating an acid soil. The bed settled rapidly the first year, but some of the original plantings still survive.

In addition to azaleas, the Lamberts brought a different approach to landscaping in Dallas. The city had been under the influence of northern nurseries, which grew such coniferous material as cedars and pines and other plants with needles. Soon a yard would be inundated with needles. The emphasis was on beds 10 to 15 feet wide in front of the home, difficult to prune and more difficult to control. Beginning with the home of Walter and Ruth Lechner, who brought them to Dallas, the Lamberts specialized in good design and execution. They believed in taste and simplicity and their ideas caught on.

Their first effort resulted in a pure explosion of color, not entirely by design. "Somehow," Henry mused, "we didn't consider the fact that the Lechner house was red when we put in red azaleas. When they bloomed, it looked like the place was on fire."

In addition to the azaleas, there were pomegranates, crabapples, camellias, wisteria, chrysanthemums and bulbs in the spring. The garden was 200 feet by 200 feet. Even the firemen would come out in fire wagons to look. It drew people from all over the state and country.

While the job was in progress, the Lechners paid another visit to Portland, Oregon, and stopped by the gardens of A.B. Lambert. Walter returned with a load of small spruce trees, a monkey tree, a ghost tree and several boxes of seed. Proudly, he informed Henry that he had purchased his cargo from Mr. Lambert for $400.

Henry raised an eyebrow. "My uncle saw you coming, didn't he?" he responded. Later, he called his uncle and said, "Please, don't sell my customers things that won't grow in Dallas."

Walter had been warned to cover the azaleas with burlap bags sewn together at the first sign of a freeze. One night, in the early hours after a long evening out, he was driving home when he heard on the radio that a freeze would reach Dallas that morning.

Tired and short-tempered, he went to bed without covering the plants. The next morning the azaleas were undamaged.

Walter adopted Joe Lambert, Sr.'s theory that a wet freeze seldom injured a plant. Once he watered the entire yard. When he woke up the next morning, every plant was covered with a light icing. Confused, he now called young Joe and explained what he had done, expressing the fear that he had probably injured the plants.

"What are you calling me for?" said Joe, sleepily. "If there's any damage, it's already been done." The next spring produced azaleas as pretty as ever.

The Lechners, as well as the Lamberts, were to know immense pleasure and pride from the azaleas they brought to Dallas. They were their gift of beauty to the city.

Chapter 9

As a man who could get things done, who had persistence and a sense of discretion, Walter Lechner found himself frequently caught up in the world of political intrigue and power.

It was never his favorite diversion; he did not have the ego or the stomach for it. But he was never reluctant to take a stand on people or issues. He was a friend to those who became famous and important in Texas politics. He was there when they needed him.

Texas governors James V Allred, Beauford Jester and Allan Shivers and Congressmen Sam Rayburn and Lyndon Johnson were among the giants of state and national government who sought his trust and support.

His first political exposure resulted from his early friendship with the young attorney, James V Allred, with whom he worked in Wichita Falls. It was a case of two war veterans meeting casually and finding an immediate bond that would grow stronger through the years.

Nearly two decades later, Allred, by then the governor of Texas, sent Walter Lechner to Mexico to represent him at the inauguration of Rodrigo Triana as governor of Coahuila. Lechner cancelled plans for a duck hunting trip to Louisiana to attend the ceremonies at Saltillo on the morning of December 1, 1937. With him, as a second representative of Governor Allred, was Matias de Llano, descendant of one of the state's pioneer families.

Lechner's first public support of Allred was in 1929 when Jimmie was elected attorney general of Texas. He was not yet in a position to help financially, but he did what he always did well—hit the streets on behalf of the candidate. In later years he contributed heavily to Allred's campaigns for governor.

In 1938, W. Lee (Pappy) O'Daniel succeeded Allred as governor of Texas when he overwhelmed a large field of candidates, including two titans of Texas politics, Colonel E.O. Thompson, the railroad commissioner, and William McCraw, the attorney general. Allred was determined that McCraw would not succeed him and put his hope in Thompson. Thompson was the power of the Texas Railroad Commission and one of the most popular vote getters in the history of the state. Not only did he have a remarkable career as a railroad commissioner, but he was a hero of World War I and later one of the most colorful and effective mayors in the history of Texas in Amarillo. But Thompson was a man of great dignity and projected an almost austere image to the public. McCraw, on the other hand, was a lively campaigner with style and a great sense of humor. He mesmerized his audiences with anecdotes and stories, but seldom spoke about the issues.

It became apparent, during the early days of the campaign, that McCraw was attracting many voters and that Thompson was losing ground. At that point Allred decided the race needed another personality to compete on the same level with McCraw.

So Allred, in an inspired piece of casting, hit upon O'Daniel, who owned the Hill Billy Flour Company and conducted a daily country music and religious radio show in Fort Worth, whose popularity was booming. His original idea was to divert votes from the light-hearted McCraw campaign and thereby benefit Thompson. Allred talked O'Daniel into running, pointing out that his radio show would enable him to reach women, a pivotal voting bloc in any state election. It was a masterful and uncanny example of a politician—Allred—who could read the public mood and who knew how to exploit it.

On radio and in person—at colorful rallies that featured his own Hill Billy band—O'Daniel worked the farm-to-market roads of Texas and forged into the big cities. It was the first total radio campaign in Texas history. Most of the newspapers ridiculed and opposed O'Daniel. At first O'Daniel had been reluctant to be a candidate, but Allred pointed out the value of the opportunity to come in close contact with his customers and to obtain a great amount of additional exposure. His campaign was imaginative in many respects. Not only did he carry along his famous radio band, but also his wife and children as part of his campaign entourage. Small barrels with slots in them were passed through the crowds for contributions of nickels, dimes and quarters. He based his whole campaign on a warm, folksy, religious approach. His crowds were tremendous from the outset since hundreds of thousands of people knew him through his radio program.

Several weeks before the end of the campaign, many realized that O'Daniel would lead the ticket and would probably win without a runoff. By this time Allred had decided that anything that kept McCraw from becoming governor was worthwhile, so on the last night of the campaign, O'Daniel broadcast his final message from the governor's mansion, with an introduction by Allred. When the votes were counted, O'Daniel had swamped the entire field of Democratic candidates, winning a heavy majority in the first primary. No second primary was necessary and the Republicans, as usual, offered only token opposition in the general election.

In the years after the First World War, in a universe as young and turbulent as the oil industry, a man on the go could get to know a lot of lawyers. Lechner did. Another one, besides Allred, was Beauford Jester, who in 1924 was handling some title work for him in Corsicana during the Wortham boom days. Again, as with Jimmie Allred, their friendship flowered and led them into the political battlefields, Walter working as usual behind the scenes.

Jester was a Texas aristocrat, son of a lawyer who became the state's lieutenant governor in 1895. He was distinguished looking, with a courtly manner and an unshakeable integrity. In the 1940's he was a member of Texas' powerful Railroad Commission which regulated the state's petroleum industry.

Early in 1946, Lechner was drilling a well in northern Louisiana when the phone rang at 2 o'clock in the morning in his hotel room. It was Jester, wanting to know how long it would take Lechner to get back to Dallas.

"Hell, Beauford, I just got here from Dallas."

"Walter, it's very necessary that I see you as soon as possible."

Lechner took a deep breath. "Well, if it's essential, I can turn around and come back to Dallas."

"It is. I've got to talk with you."

Lechner checked out of the room he and Jimmy Nowlin kept on call at the Washington Youree Hotel in Shreveport. He grabbed a bite of breakfast and by 6 a.m. was on the road back to Dallas. An hour before noon he reached the Baker Hotel in Dallas and proceeded to Suite 700 as Jester had instructed. When he walked in, he noticed that all the furniture had been removed and replaced by typewriters, desks and tables.

Jester led Walter to a closet and pulled out a box of stationery. He removed a sheet from it and handed it to his friend without expression.

It was engraved with the heading:

Jester For Governor
Walter W. Lechner
Chairman
Dallas County Campaign

Walter had been drafted to head Jester's first campaign for governor. He looked up and smiled. "Beauford, if you've gone this far and have that much confidence in me, I'll take the job."

They shook hands. "Well," said Jester, "that's one thing that's off my mind now."

Lechner placed his sister-in-law, Marie Nowlin, in charge of the office and carried the campaign through to the finish, which saw Jester defeat Dr. Homer P. Rainey, liberal former president of the University of Texas, in a solid political upset.

During the campaign, Walter encountered the lighter side of politics. The opposition occasionally harassed Marie Nowlin by telephoning the Jester headquarters in the Baker Hotel and singing a little ditty to the tune of "K-K-K-Katie"—"B-B-B-Beauford, Beautiful Beauford. . . ."

One day, Lechner asked Miss Nowlin to leave the room while he took over the phone and dealt with the telephone hecklers. When the call came, Walter was there to intercept it. He never repeated in polite company what he told them, but Marie was never bothered again by unwanted calls.

The new governor later appointed Lechner to one of the three seats on the Lake Texoma Commission, which was then attempting to work out a joint fishing agreement with the State of Oklahoma, an effort that was unsuccessful.

Walter remained a devoted supporter of Jester until that stunning Monday morning, July 11, 1949, when the governor was found dead of natural causes in a Pullman car in Houston. Jester's death in office moved Allan Shivers into the state's highest chair.

Shivers knew his predecessor had planned to appoint Lechner to the Texas Game, Fish and Oyster Commission, a position for which he had been groomed during his work on the Lake Texoma board.

On October 6, 1949, the warrant arrived, signed by Governor Allan Shivers, appointing Walter Lechner to the commission. (The name was changed to the Texas Game and Fish Commission during Lechner's term.)

Among those who rose to high places in Texas politics, Walter Lechner could be counted on to perform the jobs that were needed—the tough ones as well as the ceremonial ones.

When Thomas Dewey, then the governor of New York and later the Republican candidate for president, was in Texas to attend the funeral of his father-in-law, Shivers asked Lechner to meet him. Walter escorted Dewey from Dallas to the Oklahoma line (the funeral was held in Oklahoma) and then visited with him again before his return to New York.

At one point the governor asked Lechner where his wife was and Walter explained that she was recovering from minor surgery at Baylor Hospital in Dallas.

After seeing off the governor, Lechner went to the hospital to visit Ruth. When he walked in, she pointed proudly to a large vase of red roses with a card from Governor Dewey.

Irresistibly, Walter found himself involved in politics from the courthouse to the White House. It happened casually, almost inevitably, without his quite realizing the kind of commitment he was making. His farm in Kaufman County was in the district that sent Sam Rayburn to Congress year after year. Walter was a steady supporter of "Mr. Sam." Later, in 1940, Democratic political leader Bill Kittrell and Rayburn recruited Walter to run Lyndon Johnson's campaign for the Senate against Pappy O'Daniel in Dallas County. Walter solicited contributions and supported the future president in every way possible. But the effort was futile. The flour merchant overwhelmed the future president.

In local politics Lechner supported and worked for Smoot Schmid and Bill Decker, famous Dallas sheriffs.

After spearheading Jester's successful campaign for governor, Lechner swore—half seriously—that he was through with active politics. An ardent sportsman, he looked with pride upon his appointment to the Game, Fish and Oyster Commission, one of the most prestigious jobs in this outdoors-minded state. He considered it totally non-political and gave it his full effort and energy and even financial support at times. He knew every employee and game warden as well as his fellow commissioners.

His involvement continued, however, as he found himself in the frequent position of saying "yes" to old friends who had welcomed his support in election battles and now sought his advice and counsel in office.

At the urging of Lyndon Johnson and Sam Rayburn, he agreed to serve during World War II with the refinery division of the Petroleum Administration in Washington. He was on a fishing trip in Minnesota—the world had a curious way of calling Lechner back from his pleasures—when a telegram reached him from Harold Ickes, asking that he report to Washington, D.C. He left on the first available train. It was to him like being called into service again.

That appointment was a notable tribute to his stature within the industry but, characteristically, Lechner shrugged it off. He told his superiors he knew nothing about the refinery business. The only experience he ever had was writing bills of lading in Port Arthur when he was with Texaco in the export department.

In his time he was to see the best and the worst of the political process. The honesty of Beauford Jester made the late governor a favorite of Lechner, who saw him diligently perform his job on the Railroad Commission and his duty to the people of Texas. Jester recognized that oil was his state's most important industry and that it needed to be handled intelligently.

His father was a Democrat, but Lechner, like many modern day Texas oilmen, followed a more or less independent path in presidential politics.

Truman set in motion the attack on percentage depletion which Lechner considered absolutely essential to the continued life of independent oil explorers and producers. He was disenchanted with Eisenhower when he vetoed the Harris-Fulbright bill which would have removed price controls over natural gas from the producers. He was shocked when Johnson abandoned the oil industry as president and accelerated the war in Viet Nam. Nixon failed to live up to his support of percentage depletion for the oil industry. He also removed the control of offshore oil from state regulation and threatened to open up the valves of foreign imports when the industry announced a 25¢-per-barrel increase in crude oil.

Lechner could not accept Nixon's activities in connection with the Watergate scandals either.

Among his papers is a note bearing the seal of the White House and dated September 3, 1964, which reads:

> Dear Walter:
> Many thanks for your message extending congratulations on my nomination and good wishes for my birthday. It was kind of you to wire and I want you to know of my appreciation.
>   Sincerely,
>   Lyndon Johnson

Chapter 10

In a very real sense, the wealth that came with the East Texas discovery enabled Walter Lechner to buy the most precious commodity of all—time to enjoy the sweeter trappings of life, to sample the attractions of art and culture and travel and to become involved in the leadership of his community. He now had time to enlarge his personal horizons.

He was to do all of these things with Ruth and was to become an important national figure in the oil industry, a fighter for those principles basic to the survival of independent producers who find 80 percent of all domestic petroleum.

His oil business was at a peak, with wells in progress in Texas, Louisiana and Arkansas. But Lechner had a happy faculty of being able to separate almost totally his worlds of work and play. When he and Ruth vacationed, he mentioned business hardly at all. But when he was involved in business, he concentrated on it to the virtual exclusion of all other things.

Walter and Ruth had already traveled, in 1935, to Canada and Mexico, sharing their leisure and balancing the scales for the lean and long nights before and during Scurry County.

The year 1936 began with a towering emotional blow. On February 2, he arrived by train from a business meeting in Kansas City to find Jimmy Nowlin waiting at the station with the news that his father had died during the night. Philip Lechner was 79 when he died. He had been born at a time when most Americans scratched out a living working 12 hours a day, six or seven days a week, and when men frequently were simply ground down by exhaustion at the age of 35. He needed all his hard work and good fortune to save $300 or $400 a year from his work, his farm and his business.

In his diary that day, Walter Lechner logged one brief but eloquent notation:

"Best man in the world left it today."

Walter saw that his mother received all the proceeds of his father's insurance and arranged with the attorneys for her to continue living in the home in Terrell without pressure to sell.

In April his weight began to drop at an ominous rate, from 201 pounds to 193 in a week. He began to watch the scales daily, and the loss plainly worried him. He had developed a stomach problem that would plague him for the next 14 years. There were nights when medical science did not know a pill strong enough to ease the pain. The long siege did not really slow him down, but it certainly limited some of his old habits.

A diary entry in July records that he went to bed at 9 p.m., a sharp contrast from those entries of the 1920's when Walter hardly went to bed at all.

By late July the Lechners were planning a cruise to Australia and New Zealand, accompanied by Marie Nowlin, Ruth's older sister.

This was to be the most important trip they would ever take, notably because it brought Marie into Walter's office where she became an indispensable fixture. She was, as well, an inseparable companion to Ruth. For several years she had been a secretary with the Atlantic Oil Company.

They left Dallas at the end of July, and on August 7, off the California coast at Wilmington, boarded the U.S.S. Malola for the Southwest Pacific.

They were several days at sea, standing on the deck in the moonlight and admiring the calm Pacific and the soft ocean breezes, when Walter turned to Miss Nowlin.

"Marie," he began, "what's Atlantic paying you? No, don't tell me. Whatever it is, you just come with me and I'll pay you $150 a month more."

Right there, on the gently rolling deck of the Malola, they shook hands and the agreement was made. Marie had taken a six-months' leave of absence to join Walter and Ruth on their voyage. At the end of that time, she cleaned out her desk at Atlantic and went immediately to work for her brother-in-law.

From the beginning Marie proved an efficient, capable assistant. Lechner was able to leave a matter in her hands, with a simple "yes" or "no," and a letter would be composed and mailed without his even seeing it. He felt free of the stifling burden of office work brought on by success. He could again devote his attention to the drilling of wells or to leasing matters, secure in the knowledge that Marie would handle everything.

He eventually placed her in complete charge of his office at Eagle Mountain Lake and his personal work in Dallas.

He not only hired her away from Atlantic, but invited Marie to live with the Lechners at no expense, as a member of the family. She and Ruth were like twin sisters.

In mid-August, 1936, they docked in Honolulu, and a week later boarded the Matson Liner Monterrey bound for Sydney, Australia. The voyage and the days in Sydney and Melbourne were filled with the usual tourist delights and small complaints.

But the trip was noteworthy for one other passing moment, an uncanny bit of precognition on the part of Walter Lechner.

On August 28, 1936, he noted in his diary that "we are a gullible or careless nation. Two Japs are going over this boat, taking pictures of radio room, boat davits, etc. Sailors tell me they do it all the time."

The surprise attack at Pearl Harbor was still five years, four months and ten days in the future.

Wherever he went, Walter Lechner had an abiding curiosity about people and places and matters that touched the universe. He met strangers with ease, and they did not remain strangers long.

The Lechners and Marie Nowlin returned to the United States and to Dallas by the end of October. In December the entries in Walter's diary were once again dominated by oil talk. "I was up all night Friday running pipe on the Jones #2 well," read one log. It was the first reference to his being on an all-night vigil in some time.

In 1939 Lechner and Jimmy Nowlin drilled the discovery well in Navarro County's Bazette field. The deal was engineered by Nowlin to take over a block put together by Humble Oil & Refining Company. They drilled nine wells south of the Trinity River. Water eventually began to encroach on the field, but not before they took out over $500,000 worth of oil. They drilled three 5-million cubic feet gas wells which had to be plugged and abandoned for lack of a market. Lone Star Gas said the field was too far from Fort Worth and Dallas.

In a whimsical footnote to the venture, Ruth Lechner became president of the small company established to hold the lease. The deal was described as a farm-out from Humble, which assigned the leases to a company called Topaz Oil, which was formed to hold them. Bill Harrison named the company. "I told Ruth I wanted to make her president of it," Walter said, "so if anybody had to go on to the penitentiary, it would be her."

By this time the hub of the Lechners' social activity had become their lake home at Eagle Mountain, northwest of Fort Worth in Tarrant County. It was a retreat from the pace of inner city life and, at the same time, an ideal place to receive the family, friends and the investors and associates an independent oilman needs. But "WaRu" (from Walter and Ruth), as the lodge was named, also attracted the rich, the famous and the powerful. The lure was the gracious hospitality of the Lechners and the genuine comfort of the spacious lodge.

The lodge had sleeping accommodations for 16 people and featured a massive living room, 30 by 60 feet, with a 25-foot ceiling and a balcony completely encircling the house. The walls were of Oregon white and knotty pine.

Situated on a hill overlooking the lake, the house had a southwest exposure that kept the temperatures cool enough for blankets, even when downtown Dallas sweltered under 100-degree heat. Lechner often referred to it as his "little cabin." He bought, docked and frequently used a 33-foot cabin cruiser that was the scene of many gala social evenings and family reunions. He named it "the Maggie May."

Among those who consistently visited the lodge were the legendary Sam Rayburn, Speaker of the United States House of Representatives for so many years, and Edward J. Flynn, the national Democratic Party chief and a power under Franklin D. Roosevelt.

Government leaders, writers, artists, cartoonists, union leaders, executives, governors, senators, scientists, bankers, oilmen, generals, enlisted men, merchants, famous jazz and classical musicians, and Broadway and Hollywood stars sampled the hospitality of the house on Eagle Mountain Lake.

Frequently, the entire Henry Busse Orchestra, then usually appearing at the Cipango Club, would be hired to entertain. The elite of Dallas political and community life gathered there. The large, wood-covered guest books were filled with the distinguished names of mayors, judges, marshalls, publishers and editors. Ruth's garden and bridge clubs often met there.

On Sunday, October 13, 1940—his fiftieth birthday—Walter Lechner was host to the men and wives of the 169th Aero Squadron for their ninth reunion.

A story in the *Dallas Morning News* told how they considered themselves the luckiest men alive and had adopted 13 as their lucky number. The article goes on:

> Thirteen was chosen as the Squadron's lucky number for several reasons. First, it was organized at Love Field, December 13, 1917. When it entrained for New York to be sent overseas there were 13 coaches in the train. They spent 13 days in camp in New York in barracks Number 13, on Street 13. There were 13 ships in the convoy that took them to France. The voyage required 13 days. They spent 13 months in France and were mustered out of service on May 13, 1919. Finally, they point out that 169, the number of their outfit, is the square of 13.

The facts added up to one of those incredible coincidences that so delight students of numerology and confound those who like their explanations neat and scientific.

Lechner always considered 13 a number with personal significance as well. It was his birthdate. His office was on the thirteenth floor of the Kirby Building. And he felt that Fridays the thirteenth were especially lucky for him. He was not a generally superstitious man, but he did make it a habit always to go out the same door through which he entered any room.

Twenty-eight members of the 169th attended the 1940 reunion. Some had survived four major battles of the war—Saint Mihiel, Metz, the Argonne Forest and the Defensive Sector.

Even as the old comrades were reliving the battles of another day, the winds of war again swirled across Europe. In America, the radio had become the center of interest, an object around which families gathered in the evening to listen to reports that were consistently discouraging. Germany was again on the march.

In quick order, the King of Belgium capitulated, France surrendered, and Winston Churchill vowed that Great Britain would carry on alone, if necessary.

That summer Walter raised the American flag over WaRu Lodge. His instinct told him that the United States could not remain out of the war for long as Germany again overran her World War I allies.

Meanwhile, he began to emerge as an influence in the Texas Democratic Party and, as the national elections neared, he held frequent conversations with Sam Rayburn, Senator Tom Connally and Lyndon Johnson.

In early November he predicted a near-landslide for Roosevelt over the Republican Wendell Willkie. He spent the day of the seventh at Democratic headquarters, listening on radio to the returns that proved him correct.

Thanksgiving came and with it a feeling that the country would not have many more such peaceful holidays. Walter and Ruth spent the day in Austin, watching an underdog Texas team hand Texas A&M its first defeat in two years and knock the Aggies out of the Rose Bowl. The score was 7-0.

In March, 1941, as Yugoslavia struggled to hold out against Hitler's storm troopers, Lechner was called to Washington. His views were sought on the role of the oil industry, in the event the United States became involved in the war.

On April 9, United States Senator Morris Sheppard of Texas, a tireless champion of the state and a man Lechner admired greatly, died in Washington. A special election was scheduled to select a successor, and on April 26, Walter's diary carried the terse notation: "My candidate for U.S. Senate, Lyndon Johnson. Looks as though 8 or 10 will be in race."

He was present in Austin in May when Johnson gave the speech that opened his campaign for the Senate. He observed, "A very good speech. I am for him 100 percent on his platform—no federal oil control, depletion stays where it is, and community property tax in addition to many other things he stands for."

He was already spending much of his time in the Dallas County headquarters for Johnson, with offices in the Adolphus Hotel.

As the war drew nearer and the sense of urgency increased — the United States had ordered all its citizens out of the Far

East—Lechner became increasingly concerned over what he regarded as an indifferent or even hostile government attitude toward the oil industry. The major target of his complaint was Harold Ickes, Secretary of the Interior.

A coal strike threatened to paralyze the country that fall, and Lechner saw in it a lesson the government should not have missed. "If Mr. Ickes wants to do his duty for defense," he wrote, "he can do it not by taking over the oil industry, but by taking over and operating the coal industry. The oil industry is the only business that has been able to go ahead and operate on its own resources and is capable of delivering 100 percent in any emergency. Someone besides John L. Lewis should take over the coal industry."

He felt strongly that the men who found and produced the nation's oil needed a friendly voice in the administration, and his choice was State Senator Alvin Wirtz of Seguin. He hoped that Wirtz would be considered for the position of deputy coordinator under Ickes.

In Dallas, he was now a member of the Grand Jury, and few civic functions of any consequence were scheduled without his being represented. On October 13, 1941—his birthday—he met General Tinker of the United States Air Force and several British air marshalls at Hensley Field, Grand Prairie, Texas. The men were special guests of the Texas State Fair. Two days later he entertained Postmaster General Walker, and for the rest of the week there appeared on his schedule a procession of high ranking military figures.

Meanwhile, Lyndon Johnson had lost his race for the Senate to former governor W.L. (Pappy) O'Daniel. But as later events would prove, Johnson's career was not damaged and Walter's efforts had helped keep the race close.

On Armistice Day, November 11, 1941, Lechner felt a melancholy he could not shake. Twenty-three years before, he had been in France in the Argonne Forest. He was certain America's days of peace were numbered. He looked with grim suspicion on the talks then taking place with Japanese diplomats. "My personal opinion," he wrote on November 17, "is that the Japs are sparring for time."

Irrevocably, time and history moved toward December 7, 1941. His diary entry for that date began innocently: "Got up at

8 a.m. and went over tract of land am figuring on buying at lake." Then: "Japs broke peace today bombing Honolulu, Hawaii and Manila. My predictions of past dates seem to be coming true."

The next day President Roosevelt addressed a joint house and asked for, and received, a declaration of war. Three days later Germany and Italy officially declared war on the United States. The Lechner diary for that day predicted that "we are in for a long siege. Unless some fluke occurs this war may last five to 10 years. Not a very bright Christmas picture."

The early days of 1942 brought more shocks and reversals. The war dominated the news and the lives of those at home. Manila and Corregidor were falling and Singapore was threatened. Walter was again called to Washington to find the city in a state of turmoil and confused excitement. The Mayflower Hotel, where he usually stayed, was full, so he put his bags in Harold Young's room. Later, he moved into Sid Richardson's room and that night they were joined by another transient, Amon G. Carter of Fort Worth. Richardson was one of Texas' most important independent oilmen and Carter was publisher of the *Fort Worth Star Telegram,* the most politically influential newspaper in West Texas.

The topic of conversation was a rumored price ceiling on all oil products. Another area of concern also drew Walter Lechner's comments: "We are having labor troubles in this country again and until such time as we are able to keep down strikes, this country will never be able to lick Hitler and Japan or any other country. I fear the consequences after this war is over."

Early February marked the third year since he had taken a drink, a concession to his stomach problems which periodically flared up. In fact, his health was considerably impaired, but his schedule of activity never slowed.

On the fifteenth of that month, the day Singapore fell, he was invited to a meeting of the Petroleum Coordinating Office in Washington.

Clearly, oil would be a precious American resource and the supply of it would have a direct and dramatic effect on the course of the war. Given his stature in the industry and his general reputation, it seemed obvious that Lechner would be offered a key government position.

On March 3, a representative of the Department of the Interior offered him a post as director of the Mineral Facility Security Service. After much thought he concluded the job would not be acceptable. It would have required his staying in Washington, and he anticipated the red tape of politics would so tie his hands as to make it impossible to get the task done. He declined with regrets.

The calendar was now becoming as essential to one's daily routine as the radio. On April 9, Bataan fell. The next day Sam Rayburn delivered a speech advocating government loans to small independent oil operators. Shortly thereafter United States Senator Tom Connally of Texas followed suit.

Lechner dashed off a letter to his friend, "Mr. Sam," in which he made a forceful case for the oilman.

"This," his letter began, "is Col. E.O. Thompson's and my idea of relief for independent oilmen. The oil fields of Texas are shut 18 days this month. They have been shut 12 days for months. This is because of tanker shortages and lack of pipelines to the East.

"This will break the independent oilman unless relief is given. We give warehouse receipts for cotton and corn and wheat in storage. Oil stored in the ground is in the best warehouse on earth. The government should issue negotiable receipts to producers who are now denied the right to produce. Warehouse oil receipts should be for amounts producer would regularly be allowed to produce under proration orders, less the amount he is now allowed to produce. This will prevent the bankruptcy of many little independent producers."

By this time Walter, eager to contribute to the war effort and determined to speak up for the oil industry, had accepted an appointment to serve with the Petroleum Administration for War, Refinery Division. The agency operated in a vital area, so Lechner spent most of the next three years commuting between Texas and the nation's capital, working with John Newton of Beaumont, the chairman.

On April 16 in Dallas, Walter and Ruth served on the reception committee for Lord Halifax, the British Ambassador to the

United States, and Lady Halifax. The spirit of the elegant, formal and star-spangled affair at the Baker Hotel was summed up in a paragraph that appeared on the evening's printed program:

> Two Leaders—Two Nations—Two Peoples—united for the preservation of freedom and the rights of man throughout the world.

The war had not yet begun to turn, but now the Allies were striking back and the news was no longer a stream of discouragement. On April 18 American bombers under General Jimmy Doolittle struck Tokyo, carrying the war for the first time to the Land of the Rising Sun. "More power to the air boys," Lechner noted in his diary with the pride of a pioneer in Army aviation.

At home the oil business was in a growing crisis. Walter was convinced that a pipeline would have to be built to the East Coast before any relief could be obtained in the Southwest or mid-continent areas. He also suspected that any such move would be opposed and fought bitterly by Standard Oil and other integrated oil companies.

Everywhere the pace seemed faster, as though life were a movie run at a quick-forward speed.

On April 26, 1942, Walter heard a speech on radio by United States District Judge James V Allred, the former governor and his friend of so many years. He described the speech as "excellent" and went on to observe, "Judge is a very forceful speaker. Hope he hits the water for U.S. Senate this summer."

In his diary, he entered the fall of Tobruk to Rommel on June 22. In state politics, he predicted that Allred and O'Daniel would face each other in a run-off for the Senate.

By August the Russians had halted the German advance at the gates of Stalingrad. Lechner guessed that if the Russians did not surrender by September 15, Stalingrad would not fall.

The election results were final by Sunday, August 23, and "Pappy" Lee O'Daniel had defeated Allred by a slender margin. It was obvious that if the bandwagon voter had realized Allred's strength, he would have won.

By the first week in October the war had moved a little nearer to home. Walter's only son, Rembert, was inducted into the Army.

Later that month Walter was commissioned a captain in the United States Coast Guard Auxiliary and placed in charge of the Dallas and Fort Worth flotillas. Immediately, he began insulating and converting his cabin cruiser for duty as a patrol boat.

Another Armistice Day passed on November 11, and this time Walter noted: "What a day with the world on fire again."

In a meeting at the Adolphus Hotel, Congressman Wright Patman asked him to prepare some data for his House Banking and Currency Committee on the price of crude and the oil outlook in general and to appear before that body in December. Walter agreed.

The first anniversary of Pearl Harbor was a time for reflection, and Walter could not ignore a statement by Tojo, the Japanese War Lord, that America must be annihilated. "He doesn't know what he is undertaking," he wrote. "We have just begun to get mad."

Within a few days the letter Wright Patman had requested from him was presented to Congress. In it, Walter Lechner advocated increasing the price of crude substantially so independent wildcatters could continue operations and help meet the increasing demand. He pointed out that labor, material and service costs, as well as taxes, were increasing by leaps and bounds while crude oil prices remained under strict controls at $1.20 per barrel. He said 80 to 85 percent of new discoveries had been made by independent wildcatters, "so you can readily understand why it is so vitally important to keep them going." His report included statistics on costs, demand and other data.

Lechner discouraged the idea of a government subsidy. "A government subsidy has been suggested as a way to handle future exploration. This method of discovering new reserves would be impractical and fraught with dangers from the start," he said.

In a brief attached to the letter, he discussed stripper wells, with facts and figures for the State of Texas in 1941:

> The effect of premature abandonment of stripper wells may readily be appreciated by the fact that the average stripper well abandoned within the past four years still had a reserve of 10,407 barrels of unproduced oil. On the basis of reports received from

stripper wells over the past four years, low crude prices have resulted in the loss of over 85 million barrels of unproduced oil remaining in the underground reservoir.

In the first half of 1943 the war news became more favorable, but Lechner's irritation with government ineptness and bureaucracy continued to mount.

On the last day of April the coal miners were striking, and Walter had some unpleasant remarks about the character of John L. Lewis. He listened to a radio speech by President Roosevelt, deploring the action of the miners, and he wrote in his diary: "It seems unfair to me that the miners should be given all the consideration in the world and the oilmen are only sideswiped by a bunch in Washington that want to see the little man collapse."

His diary entry on May 3 predicted that the Allies would attempt to invade the continent of Europe by July 1, 1943.

His thoughts and energy were now almost evenly divided between the war and the oil industry. In late June he telephoned Sam Rayburn and repeated his position that the industry did not want a government subsidy.

During the first week of July he noted, with satisfaction, that the Allies had invaded Sicily. The big push had begun.

The war had definitely turned in favor of the Allies. If the end were not yet in sight, the outcome at least seemed clear to Walter Lechner. On July 25 rumors were afloat that Benito Mussolini had fled from Italy—or was attempting to do so. Walter was confident Italy would be out of the war by mid-September.

His ability to foresee trends, to anticipate future trouble spots—borne out so often by later developments—was still active. "I may be dumb as an ox," he noted, "but someone at the head of our oil industry is a damn fool in approving so much oil to be shipped to Spain. Franco, in my opinion, is bypassing it on to Adolph." All the thoughts were recorded in his diary or in letters to friends in Congress.

Still in July he left Dallas and drove to Bonham for an hour session with Sam Rayburn about the price increase in oil. "It is beyond me," the Speaker told him, "why the powers-that-be don't grant an increase as recommended." Walter agreed and added: "Labor, supply, equipment and service costs are apparently uncontrolled. Someone is obviously out to annihilate the small oil producer."

He stayed in almost daily touch with Rayburn regarding the thorny issue of realistic prices for oil. In the meantime, he was constantly trying to increase production in his wells and searching for new oil. His principal interest was still oil production for the war effort.

His relaxation centered around the home at WaRu on Eagle Mountain Lake and there, due to the war-imposed shortage of labor, he did much of the work and many odd jobs himself.

The frustrations of the oil industry were increasing. At one point, Lechner noted in his diary: "If Ickes can cut the amount of gas used in the U.S., it seems to me he can cause an increase in price. Looks screwy to me." On August 13, the Secretary of the Interior announced that gasoline ration cards would have only three gallons of value. Lechner entered the news in his diary and alongside it placed a cryptic comment.

Later, he added, "If Mr. Ickes can curtail gasoline, why in the hell with all of his 'God-given (FDR) powers' can't he make his price increase stick. Standard Oil of New Jersey not quite ready, I presume." Walter had little confidence in "the Octopus." He was irritated at seeing many of his independent oil producer friends going broke and selling out to majors every day.

Walter was becoming a keen student of the nuances of the war bulletins. When he heard one night that Berlin radio was off the air, he knew it was a sure sign that the Royal Air Force had conducted another bombing raid. He analyzed the stories behind the headlines and listened to his sources in Washington and formed his own opinions.

He grew increasingly wary of Russia's attitude toward the Allies and just as disturbed by the expansion of big government, partly concealed by the exigencies of war.

"Am prone to think we will have plenty of deep thinking and work to do in this country," he wrote, "unless our bureaucratic form of government is revised. Personally, I'm for states' rights 100 percent."

He was merciless in his appraisal of the men who executed the nation's policies. "The bureaucrats in Washington," he declared, "especially the OPA, are to blame for everything that happens to the oil business. They have stymied every effort to increase the price of crude and create an incentive for more drilling of wildcat wells."

He developed a hatred for Leon Henderson, the price czar, and Ickes.

His gathering despair was broken by the favorable tidings from Europe. Italy fell on September 8, fulfilling his prediction made some weeks before. But the country was still to be occupied, and American troops moved on Naples and Rome.

The war put a major crimp in the annual reunion of the 169th Aero Squadron. They met in Dallas on October 16 and a newspaper clipping refers to a "somewhat forlorn little group. Of the 300 brisk young fliers who served, only about 175 still are living and only those within the radius of a few miles could attend." But the bottle of vin blanc which the men of the 169th had chipped in for in 1919, to be drunk by the last survivor, was still a long way from being opened.

Chapter 11

TIPRO

As World War II ended, the complexion of the independent oil-producing industry was changing rapidly, but price controls on crude oil still remained at an unrealistically low level.

At the same time the major companies were cooking up a scheme with the Roosevelt administration under the name of the Anglo-American Oil Treaty. This was the outgrowth of a close relationship between American and British oil leaders during World War II. The purpose of the legislation, which had passed through Congress at one time only to be stymied before it got to the President, was never quite clear. It was being sold as a package to assure intelligent development of foreign oil fields in the sphere of American and British companies.

Harry Sinclair of Sinclair Oil Company was the first to fight the plan. Then came scattered independents, who called it an attempt to create an oil cartel of large international companies, especially those companies which would later be called "The Seven Sisters."

Texas oilmen, especially Glenn McCarthy, the fabulous Houston wildcatter, took up the battle and decided to organize to combat the legislation.

Independents were also plagued by state problems in the areas of conservation, taxation and regulation. To deal with these diverse problems, the independents decided to form an organization in Texas under the name of the Texas Independent Producers and Royalty Owners Association, to become known as TIPRO.

On March 1, 1946, a group of about 60 independents and royalty owners from all over the state gathered to hold their first meeting. It was held in Austin with Walter and twenty other men forming the organizing committee.

Walter was sent to Washington to see what Sam Rayburn thought of the idea and to learn whether there would be any roadblocks. Rayburn, whose sympathy with the oil industry independents was based on his recognition of their tremendous value to the general public and their direct benefits to his own state, gave the idea his full blessing.

TIPRO was then formally organized on April 18, 1946. The voice of the small oil operator, in an era when specialization and centralization is threatening to make individualism an obsolete pattern in American life, was now to be heard.

The founding president was H.J. "Jack" Porter of Houston. He later became an almost-successful candidate for the U.S. Senate on the Republican ticket. Walter was appointed to the executive committee and his old Scurry County sidekick, E.I. "Tommy" Thompson, was the first executive vice president. Walter, along with his friends, was a moving power in its early history. It became a most active and influential association and remains so.

Through the help of Rayburn and Senator Tom Connally, chairman of the Foreign Relations Committee, and a group of dedicated Republicans, the Anglo-American treaty was set aside and never became law. Soon also price controls on crude oil were removed. Free enterprise seemed to have thwarted two prime enemies, largely through the work of organized Texas independent oilmen.

SCURRY BOOMS

Walter and Ruth often thought about the days in Snyder when they each worked for a salary and a nebulous interest in the profits of an oil company which was never realized.

Those days were a struggle, but they were also some of the happiest and most rewarding days of their lives. Walter kept up with developments in West Texas, especially Scurry County. He had always maintained that when equipment was capable of going deeper than the shallow oil he had discovered there, much richer reservoirs of oil would be found. Often, since he had made his fortune, he had considered going back and looking into the situation himself now that deep drilling equipment was available.

He knew, for instance, that Humble Oil and Refining Company was testing the area. He was certain that great company would find deeper oil. In August, 1948, he was surprised to read in the oil news that Humble had decided to abandon its Scurry County search and had given up its leases in the area. The fact is, an Humble well had been drilled right through the Canyon Reef, one of the greatest oil storehouses in history, without detecting the magnificent treasure.

Walter picked up the telephone and called his old friend, Cub Murphy, who had helped him assemble his drilling block back in the twenties. Cub said he believed their mutual friend, C. V. Thompson, had received a relatively large lease back from Humble just as he was hoping the company would renew it. Walter told him to go get it at once and to pay whatever Thompson wanted. Thompson wanted $5 an acre, which both Cub and Walter considered reasonable, so Cub took the lease. It was about 640 acres, or a section of land, for $3,200. Acquisition men for other companies were already in the field rounding up leases, but Thompson preferred to do business with Murphy and Lechner, men he knew. In fact, he was certain Walter would never leave his land untested.

What was to be rated later as the discovery of the fabulous Canyon Reef oil field was drilled by Sun Oil Company and Humble in a joint venture well in August on their Schattel lease. But this was to remain a one-well pool on the edge of the great field. The companies didn't realize what they had found.

In the meantime Walter had obtained his lease. He sold 80 acres for $25,000 and another 40 acres for even more as the play became hot in Scurry. He had also offered his associate, Ray Hubbard, a chance to purchase half interest in the lease at the $5-an-acre price. Walter told Hubbard they would probably drill the lease themselves. After delaying a day or so, possibly to look into the matter or to arrange financing, Hubbard came in.

Then on November 5, 1948, Magnolia Petroleum Company discovered the Kelly area of the field on its Winston lease on the extreme southeast edge of the thick body of oil. About 12 miles north and slightly east of that well, Standard of Texas brought in the first major well in the field on its Brown lease on November 21.

Six weeks later, on January 6, 1949, Lion Oil and Refining Company hit the jackpot on the C. T. McLaughlin ranch, opening up the famous Diamond "M" field, named after the ranch. That extended the producing area some 35 miles from end to end. Other areas opened later.

Walter was getting ready to drill after sending Ruth's nephew, J.H. Bartlett, a young geologist who had served in the navy in World War II, to check surface and other conditions on his acreage. Then he started scouting for a good rig to begin operations. He couldn't miss. He was between producing wells. His friend, R. E. "Bob" Smith of Houston, one of the wealthiest, luckiest and most colorful oilmen in Texas history, surrounded him with leases.

It was in this atmosphere one Sunday morning at Eagle Mountain Lodge as Walter was drinking coffee and poring over his notes on Scurry that a stranger from Midland rang his doorbell.

He was an acquisition man from Shell Oil Company, late as usual on a big field but determined to get in at any price. He wanted to lease the C. V. Thompson tract. Walter informed him that it was not his intention to lease the property since he planned to drill it himself. The man said Shell was determined to get acreage in the field and would pay a premium price. In fact he offered $500 an acre, about $250,000, plus a sixth royalty.

Walter informed the Shell man that if he were going to lease the property, it would have to be much more than that. Asked what it would take, Walter told him $1,000 an acre, almost $520,000 and a half royalty. That, his visitor said, was preposterous.

"Then, forget it," Walter said. "I told you I was going to drill it myself, anyway. Sit down. Have a cup of coffee and let's talk about something else."

The man excused himself and said he wanted to call Shell's Houston office.

"On Sunday?" Walter asked.

"Yes, I told you they want in on Scurry County and they are sitting there waiting for me to deliver your lease to them."

"I wish I could accommodate you, but I can't at your price. . ."

The Shell man went into another room and called Houston. When he came back, he said Shell would sweeten the pot, but not for as much as Walter wanted.

"That's O.K.," Walter told him. "Drink your coffee and then call them back and say you pooped out."

Walter felt certain Shell would not meet his price. In fact, had he thought they would, he would have gone higher. He actually did want to drill the tract himself despite the high cost of drilling a deep hole in Scurry.

The man drank his coffee. Before he was through, the phone rang. The Shell Houston office wanted to talk to the acquisition man. He answered and explained Walter was serious and would go no lower under any circumstance and seemed happy that Shell would not pay the price. He came back, sat down and poured himself another cup of coffee. About halfway through it, he said, "Mr. Lechner, Houston says to pay you your price. Here, fill in the figures and sign this. I think I am some kind of hero for getting anything in Scurry."

It was the best oil deal Walter ever made. As soon as the first well came in with more than 700 feet of solid oil sand at the Canyon Reef at a depth of 6,500 feet, it was obvious to Walter, Ray and Bartlett that the whole Thompson lease would produce.

He and Ruth then decided to give half of their share of the royalty to Walter's son, Rembert, and to their brothers and sisters. There were five on each side of the family. The other half was retained by Walter. From that time on, each of the brothers and sisters and Rembert received an income of about $1,000 a month. As they died, the remainder of the royalty was divided equally among those remaining. Walter's sister, Lucille, who still lived alone at the farmhouse in Terrell, and Rembert were receiving the entire Lechner quarter of the division in 1977. Ruth still had a sister, a brother, and a niece to divide the other quarter.

Of course, this was an unexpected bonanza for Ray Hubbard. At that time he and Walter were on the last legs of their 50-50 agreement from early East Texas days. That is why Walter offered him the half interest in the deal. If Ray had financial troubles then, they were over for life with a bonus of $260,000 and an income of almost $20,000 monthly for his share. Shell had drilled 12 wells on the tract under a 40-acre spacing pattern. When Walter and Ray ended their old agreement, each had the royalty from six wells.

In 1977 the wells were producing as good as they were the day they came in. The field had been unitized. There were about 2,000 wells in the gigantic field which contained some four billion barrels of oil. In 1951 engineers reported that only 2 percent of the oil could be recovered in competitive operations, but that recovery could be doubled by pressure maintenance if the field were unitized and placed under a single operator.

The successful formation of Diamond "M," Sacroc, Sharon Ridge, and the Cogdell units, and the initiation of fluid injection, doubled the field's production potential. Future developments could again double its potential production. But even under present conditions it is anticipated that the Canyon Reef field will flow almost unabated for another 70 years, the exact anticipated life of the East Texas field.

There was a sidelight to the field that Walter tells with some amusement, mixed with a noticeable tinge of regret. He and his good friend, Leon English, had always contended that deeper oil would be found when the tools and techniques were available to reach the depths.

One day after he had made his fortune in East Texas, a friend, Allen Warren, grew weary of his sheep ranch. He offered it to Walter for $7 an acre. There were about 5,000 acres on the ranch. Walter thought it over, but, as in the Taylor Lee case in East Texas, he decided he was too involved in the war and other efforts to take on a sheep ranch. He told Warren to wait until the war was over. Then, however, it was too late. In the meantime C. T. McLaughlin bought the ranch and named it the Diamond "M." It turned out to be possibly the single most valuable piece of oil real estate in the nation.

McLaughlin was an old friend of Walter's. They both had been in the flying service in World War I. Both returned afterwards to get their feet wet in the oil business in Burkburnett and Wichita Falls. Both had made and preserved fortunes through determination and hard work, plus that illusive mite of good luck. And both had always been fascinated by deep oil's chances in Scurry County.

"Mr. Mac" McLaughlin died in the middle seventies one of the wealthiest and most respected men in Texas oil history. And still one of Walter's best friends.

Walter had no regrets over McLaughlin's good fortune at Diamond "M," but he insists even today that he is probably the only man in the world who twice turned down a billion dollars that were offered to him for almost nothing.

THE OFFICE

Even though some of Walter's most important business transactions occurred in such places as coffee shops, street corners, hotels and his home, the distinguished Kirby Building has been his official place of business since 1930. Those were the days when he drove a Model A Ford convertible coupe which he used as his field car. His life has gone through many roles since then. From lessor (one who always tried to please the lessee) to working on the rigs day and night to successful oil operator.

In 1973, Earline Shelton showed up in his office to apply for work. Harry Swartz, Lechner's office manager for about 25 years, had told her about a vacancy for a secretarial job. This was a remarkable stroke of luck for Walter. Earline had spent 12 years in Washington as an executive secretary for a string of Republican congressmen, most of whom had been defeated. She finally decided Washington was not her cup of tea.

Earline was the nearest thing possible to Marie. In a few weeks she was able to attend to any phase of his office work, in addition to being a perfect secretary. Again he could tell her what he wanted to say and to whom a letter was to be sent and that was sufficient dictation. She kept his office records and bank account and performed the various duties connected with the oil-producing business. After Harry Swartz retired because of ill health, Earline was made Walter's office manager, January 1, 1974.

Despite his age, past serious illnesses and dimming eyesight, Walter was in his office for several hours almost daily. He still shared business interests with Jimmy Nowlin, who was keeping several wells producing on a stripper basis near Coleman, and was always on the lookout for other deals. In the past few years, Walter had attempted to lease land near Glen Rose and had even had his old crony, Leon English, doing hard rock geological work in that and other areas. His interest in finding new oil and gas, which he knew the country desperately needed, had not flagged.

His other veteran, Paul Barr, was still only a hoot and a holler away on Barr's restful place just outside Dallas.

A MEMORABLE JOURNEY

The day was bright, sparkling, cool and crisp. A few scattered cumulus clouds drifted overhead in the blue Texas sky.

Samuel Beale, Walter's right-hand man and driver, brought the Lechners' car to the front door of the Maple Terrace House where the Lechners resided in a well-decorated and spacious two-bedroom apartment. Sam was a large, rangy black man who knew more about football and baseball and driving an automobile than most people.

Walter climbed in the front with Sam, as he always did. A friend sat in the rear seat. Then they took off on a memorable trip.

Their main destination was the J.M. Tuttle lease in Gregg County, the Humack Oil Company's field headquarters. But the first stop was to be Terrell, about 31 miles east of Dallas, where Walter had spent his childhood and young manhood.

Terrell is a clean, progressive small town. It is surrounded by fields of fine cotton land. The landscape is slightly rolling with occasional large clumps of water oaks. On the edge of town is the old Lechner farm where Walter's youngest sister, Lucille, lives.

On the outskirts of Terrell is a plot of land where a number of British flyers were buried after being killed in training during World War II. It was to visit those graves and to pay tribute to those young heroes that The Right Honorable Viscount Halifax and Lady Halifax had visited Terrell in company with Walter and Ruth and other Dallas notables in 1942. At the time Lord Halifax was the British Ambassador to the United States. Lord Halifax and several British air officers had been honored with a formal dinner at the Baker Hotel, attended by Ruth and Walter. Walter sat at the head table with the mayor of Dallas, distinguished community leaders and their ladies.

Along the highway to Longview, Walter spoke of the tragedy of Eagle Mountain Lake when Marie, Ruth's beloved sister and Walter's secretary, became ill. Ruth and Marie had grown extremely close throughout the years and formed a sisterly bondage of respect and affection. Where you saw one you saw the other.

Marie was ill for quite awhile before passing away on June 1, 1964. "It really hit me a blow—right between the eyes," an instantaneous memory was reflected in Walter's eyes as he spoke these words softly, while expressing deep emotion.

They didn't go to the lake anymore after that. Marie's spirit was gone from the house and only memories of the good times spent when they all three shared the lodge remained. A year later it was sold. After leaving WaRu, the Lechners moved into the Melrose Apartments for a few years, then into another apartment at the Maple Terrace House.

When he started the lodge, Walter told Paul Barr, his accountant who had been with him since the roaring days of East Texas, to record every nickle the land and structure cost. One day when work was nearing completion, he asked Paul if he were keeping the record.

"I have it right down to the last penny," Paul said and started to go on. But Walter threw up his hand and shouted for him to stop.

"No, no!" Walter cried. "Don't tell me. I just wanted to be certain you had the information."

Paul said the Scotch had a way of coming out in Walter every now and then.

"He knew if I told him, he might stop before the lodge was completed," Paul said later.

Walter explained that one reason he built WaRu was his poor health. He had started losing weight for no obvious reason. He dropped from 202 pounds down to around 158 over a year or so. Doctors diagnosed his condition as duodenal ulcers. Two internal medicine specialists treated him for 14 years. They took him off alcohol, tobacco and rich food.

A few years after the lodge was sold, the doctors came to the Lechner apartment and prescribed an even more rigorous diet and other sacrifices. Ruth rebelled and fired them both. She told Walter all he needed was to start taking a drink now and then, smoking a few cigars and eating like a human being. It was a prescription Walter liked and it worked. A few months later, he was almost back to normal.

Then they started going to their present physician, Dr. R.K. Bass.

As the car whirred along the highway, Walter recalled the golden anniversary party he and Ruth enjoyed on September 1,

1970, at the Brook Hollow Country Club. It was a grand affair with about 600 of their friends coming from throughout the state of Texas and other places like New Orleans to help celebrate this memorable occasion. The festivities began around 6:30 that night. Champagne was flowing, with food galore and music and dancing until midnight.

Violins played as Walter and Ruth started off the dancing to the Anniversary Waltz. The fifty years they had shared together were beautifully portrayed as they waltzed across the dance floor and looked upon all the faces that had gathered for this reunion of their union.

In the middle seventies, before he had two heart attacks and an eye operation, Walter constantly smoked large, expensive (but not imported) cigars. He carried a cane for support due to a knee injury suffered on a hunting trip. He always wore a semi-broad brimmed black hat and dark suits. His favorite dining places were one of the clubs he helped found and the private club at the Baker Hotel. There a man could get a drink before the state repealed prohibition of liquor by the drink.

He was still a vigorous man of some 180 pounds and about 5 feet and 9 inches tall. He liked the outdoors and spent considerable time on his farm at Terrell and in the field looking at oil prospects, usually with Jimmy Nowlin.

On the highway to Longview wherever the car stopped, at cafes or service stations, everyone knew and greeted Walter. He never forgot a name, a face or an event. His recall was perfect and his mind, deep into his eighties, was as active and clear as ever.

After a while the car passed Van and he pointed out the road to the oil field that really brought the flurry of interest in all East Texas prospects. Going through the Gladewater section, he pointed out a small stream, the Sabine River. It went down to Sabine Pass where it emptied into the Gulf of Mexico. It was also the geological feature that formed the uplift to provide the trap for the Woodbine sand on its eastern edge. Walter waved his arms in a wide swing and said that was the vast acreage Taylor Lee had offered him. It was now a sea of oil wells.

"Woodbine," he explained, "is the name of vines that bind themselves around trees. One of these is the wild honeysuckle."

The car veered north on a winding country road toward the Tuttle lease where Walter maintained his small center of operations. Beautiful houses lined the road, nestled among tall pines, over the hills and down the valleys. Walter's heart seemed to beat a little faster as old memories and old stories came to his mind.

The derricks were small compared with those in the Gulf Coast and West Texas. That was because the wells were only from half to a third or less as deep as those wells.

They crossed streams over cattle guards. Finally the car stopped in what looked like an oil field junk yard. It was located on level ground on the edge of a deep ravine that separated the area from a towering hill just opposite Walter's pump house and warehouse. He inspected the wells and the warehouses and talked with his field man. Then he recounted how he and Barney Skipper had traveled this country, mostly by foot, to line up leases.

Then they went to the McGrede lease, his favorite, where he operated 14 oil wells owned jointly by him and Sohio Oil Company. He said every oil company in the field, and some from the outside, were trying to buy his wells, but he wasn't selling any more.

On the extreme east side of the field, the two men came to the McKinley lease, which was once considered a likely spot to drill the first Farrell-Moncrief well. It was turned down, he said, because the Lathrop tract was in the center of the Skipper-Lechner 9,300 acres.

It was fortunate it was turned down because wells on the tract were not commercial and would have dampened enthusiasm for drilling in the rich north half of the field for years. That, he said, would have been the end of Farrell and Moncrief because they were in the field on a shoestring. He said he and Skipper also would probably have run out of time, money and hope. In later years, however, a rich gas sand had been found much deeper on the McKinley tract. But even if the gas had been found in 1931, there was no market for it.

Walter also pointed out the acreage Diamond Joe Reynolds had offered for $10 an acre. It was barren of wells. A deal with Reynolds would have been even more devastating to the hopes of Farrell, Moncrief, Skipper and Lechner than the McKinley lease. People say luck often plays a part in oil finding.

The next stop was the gas field Skipper operated near Longview. He begged oilmen and investors to drill it. Finally he had enough money to drill it himself, so with gas at around 50 cents, Barney made another fortune.

The last stop in the field was the Lathrop well. It was still an emotional sight for Walter. And it was still flowing at a good rate. The computers reported that the well would be producing another 30, maybe 40, years. It had already given up almost a half million barrels of oil.

Recently there had been sad news from Upshur, the most northerly county in the great field. The J.D. Richardson No. 1 well, drilled by the Mudge Oil Co., the first of the county's 4,000 wells, was plugged and abandoned after 45 years of production.

On the way to the Gregg Hotel, Walter and his friend passed over the old L&H Pipeline. It was the line Lechner and Hubbard operated to move their oil out of the field. He had sold it to Al Meadows of the General American Oil Company.

That night at the Gregg Hotel, Walter talked of many things. One was his stint on the Game, Fish and Oyster Commission. That was the job Governor Shivers had appointed him to because he knew Beauford Jester intended to do so before his tragic heart attack.

Walter knew every warden by his first name. He was responsible for equipping them all with two-way radios so they could cover far more territory. In all his days on the commission, he never accepted his stipend and always paid his own expenses.

The next day the two men called on Mrs. Barney Skipper at her spacious and delightful home on the highway between Longview and Kilgore. It was on a low hill covered with typical East Texas oak and pine.

Barney, Mary Skipper explained, had intended to build a cottage. But she went to New Orleans and saw two or three magnificent chandeliers, one that had been used by an aristocratic German family, a platter owned by the Pullman family and other pieces formerly owned by the Rothschilds. A fireplace owned by a German nobleman who had often entertained Beethoven was among other treasures she found. She went back to the hotel to get Barney and took him to the French quarter to look at her finds. He also liked them and bought them. Then they had to double or triple the size of the house.

Walter and his friend continued on to Harrison County to look at the spot where Dr. Hugh Tucker had stood that day so many years ago, picked up the small piece of lignite and predicted that some day the greatest oil field ever found would be discovered about 50 miles west of there.

The last stop on the trip was a gas well Walter and Roger Lacy had drilled in the Harleton area. It was reported the well came in making more than 350 million cubic feet of gas a day. Lacy made a deal with Arkansas Natural Gas Co. in Shreveport to take the gas. It was offered to Lone Star Gas Company in Dallas, but was turned down because it was too far from Dallas.

PERSPECTIVE ON EAST TEXAS

Walter took a backward look at the great East Texas field in 1976. Its statistics were still almost beyond imagination.

The *Kilgore News* reported facts from several sources to show the enormity of the great black giant of American oil. Through 1976 the field had produced 4.3 billion barrels of the finest oil outside Pennsylvania. It was estimated to contain at least another 2.5 billion barrels of crude.

In 1975 it produced 74,806,797 barrels from 12,902 wells. Its least productive year was 1964 when it gave up only 40,021,909 barrels. That was in the midst of the nation's great unnecessary importation of foreign crude which chased more than 30,000 in-dependent explorers out of the business. Its year of greatest production was 1933 when 11,875 wells flowed 204,954,000 bar-rels of oil. That was when Walter was in his heyday. Although probably as many as 35,000 or possibly 40,000 wells had been drilled in the field, the year of the greatest number of producing wells was 1939 when there were 25,997 wells flowing. There were even more dry holes, abandoned wells and tests too far off the perimeter of the field.

Unfortunately for the old field, in 1975 many areas, especially those around Kilgore, had ceased flowing due to the encroach-ment of salt water into the Woodbine sand. In that year only 5,585 wells were flowing and the other 7,317 wells were pumping. Most of Walter's wells, however, were still flowing. The majors were trying to buy him out, but he wasn't selling.

Harold M. Smotherman, vice president of Tyler's Citizens First National Bank and a petroleum engineer, predicted that it would take operators another 60 years to get to the last barrel of East Texas oil, carrying the old field's life well into the twenty-first century.

Smotherman said that unfortunately most of the field's oil was selling at $5.25, as "old oil," rather than at the high prices of about $12 a barrel. Yet it was oil second only to Pennsylvania crude in quality and true value. But, of course, the same fate had befallen the old Pennsylvania wells. This, Walter thought, was a great example of the ignorance of bureaucrats. This very policy, he said, had chased the independents out of the field of exploration and made the majors, who found no more than 10 to 15 percent of the nation's oil, richer than ever. The majors were carrying on a campaign to buy out independents all over the country, but still Walter resisted.

As one rides through the rolling East Texas countryside from Tyler on the northwest to Henderson on the southwest, it is gratifying to see the successful results of this daring experiment in petroleum conservation and "university" of effective petroleum technology, especially the knowledge of reservoir mechanics. In their defense, it must be added this was largely due to the major companies who had promoted and financed the fight for effective conservation, often with the help of solid independents such as Walter.

Here were beautiful homes and delightful countryside. The towns and small cities were clean, prosperous and progressive. Before the field came in, this was a dreary, poverty-stricken, hopeless part of Texas.

Now it was blessed not only with fine communities, but with good schools, lovely churches, strong colleges and thriving farms. Had not conservation and engineering been employed (largely under the driving power of the late General Ernest O. Thompson and Governors Allred and Sterling), this would have been just another devastated and abandoned boom town on a tremendous scale. Also, chances are that no more than half a billion barrels of oil, if that much, would ever have been produced. The six billion barrels would have been gobbled up by salt water.

The lessons of conservation and engineering were soon applied to the rest of the state, the nation and the world. Knowledge gained here was exported to Venezuela, the Middle East and even the Soviet Union. It is now being used in Alaska, the North Sea and the offshore areas of the world.

The historical significance of this magnificent field was truly as great as that of Spindletop in another part of East Texas that in 1901 ushered in the age of liquid fuel, moving petroleum out of the era when oil was used primarily for illumination and lubrication.

Walter looked back with due pride to his part in bringing about the realization of this great blanket of Woodbine oil sand.

RECOLLECTIONS

As Walter entered 1977, he was in Presbyterian Hospital in Dallas with heart trouble for the third time in two years. Yet his spirit and remarkable physique stood him in good stead another time despite his being in his eighty-seventh year. He was up and about in six weeks.

The early oilmen are a distinct breed, having known the endurance of many hours spent laboriously around rigs. After maybe four hours of sleep, Walter would be out working on a well at such hours as 4:00 in the morning. This is the spirit of a pioneer oilman who brought in some of the very first major producing wells in the United States.

Walter never had any particular trouble handling the money he made finding oil. He might sometimes worry about the fortunes he lost in dry holes trying to find more oil. He could not help, however, pondering the contention of Barney Skipper that East Texas was a much happier place before wealth came so suddenly to many of his friends and neighbors.

Then in 1965 he read a story in the *Fort Worth Press* about his friend, Schields L. Fowler, then 97. Fowler regretted having brought in the famous "Fowler's Folly" that opened the rich Burkburnett field and led to the discovery of several other bonanzas in North Texas in 1918, just before Walter came home from World War I.

Fowler said that, to that very day, he wished he hadn't drilled that discovery well in Burkburnett. He said there was happiness in struggle and especially in cattle ranching. After he got $1.8

million from Magnolia Petroleum Company for his share of the company's stock, plus valuable royalty, he was rich. He said he never had a truly happy day after that.

Walter thought about the stories of Skipper and Fowler, but he didn't understand them. He thought money could bring happiness if it were used right. He had used his right. He had made everyone in his family, and that included Ruth's family, virtually independent. He felt he had helped many farmers to become more financially independent after years of drudgery and near starvation. He had done his best to help his city, state and nation.

Most of all, he reasoned, he had made a host of fine friends in the greatest industry on earth and he had helped in a small way provide the fuel that made his nation and its people more prosperous. It had helped provide them with the world's highest standard of living and total freedom from despots and tyrants. That, he thought, was nothing about which to have many regrets.

What he regretted most in 1977 was that his own government seemed to be determined to force nationalism on the people. He abhorred corrupt, self-serving and irresponsible antics of Congress and Presidents Nixon, Ford and Carter toward the energy situation. He saw no reason for requiring the public to make unnecessary sacrifices simply to enable the government to "chastise" the petroleum industry for its success in helping make the United States the leading nation in the world in agriculture, science, education and industry.

He knew from his long years of experience that the hidden supplies of oil and gas were sufficient for at least another century. Some experts had even said that the country had enough natural gas to last from 1,000 to 2,500 years and that it possessed coal supplies for hundreds of years. Shale oil was even more plentiful. These were all available for use in unlimited amounts as necessary research was conducted on nuclear, geothermal, solar and other exotic forms of energy.

"Why should the people sacrifice needlessly because politicians are ignorant?" he asked after President Carter's energy plan was announced in April, 1977.

The only things preventing these expectations, he said, were the government itself and a president who would draw up energy plans to force regimentation on the people without inviting a single experienced oil, natural gas, coal or shale oil man to offer suggestions for more energy instead of total dependence on conservation at great pain.

Even as he approached his ninetieth birthday after three serious heart attacks, near blindness and other disabilities of his age, he was still going to the thirteenth floor office in downtown Dallas to put in more work in a four-hour day than the usual bureaucrat puts in in a week's time.

His love for his old Democratic party was virtually gone and his attraction to the Republican party was still not strong. He maintained a high respect for such men as John Connally, Allan Shivers and a handful of other strong Texans whose intelligent devotion to the public interest was still "alive and kicking," as he put it. He wished there were more Allreds, Jesters, Rayburns and Trumans around.

Walter had remained out of the news columns. He still had an army of old time friends, although some such as Manuel "Lone Wolf" Gonzaullas, Texas Ranger hero of Kilgore, had died during the year.

"He was a gentleman in some respects, but he was a law enforcement officer in other respects," Walter commented. "He had the clearest blue eyes I ever saw in a man's head and he looked you right in the eye when he was talking to you."

He had a natural yearning for the old days, especially those in East Texas where he achieved his most important goals by work and the ability to lay out a plan and endure the difficulties necessary to its success. But he also thought often of Terrell, then his first job in Dallas and his earlier days in East Texas, Port Arthur and the salt dome fields, where he got his baptism of fire as an oilman. He also dwelt on the war and the great North Central Texas booms at Burkburnett, Ranger and Desdemona. His finishing school had been Scurry County where he did everything an oilman had to do from the derrick floor to the office.

Yet, looking back, East Texas and his association with Barney Skipper and the others had to be the highlight of a career.

Then his thoughts turned to his wonderful wife, Ruth, who was with him mentally every day from the time they were married in Dallas. In all his reminiscing he could think only of friends. If he had made an enemy, he didn't recall it. He hoped he had not.

He would always look back on one day in the seventies when he met and visited for the first time with H.L. Hunt in Hunt's Dallas office. They recalled days in East Texas when they didn't even know one another, working at opposite ends of the field. And they munched on one of those natural food lunches Hunt often brought to the office in a manila bag. There was hardly a big man in oil that Walter didn't know now.

Considering all these things, he thought to himself no man ever lived a better life.

Index

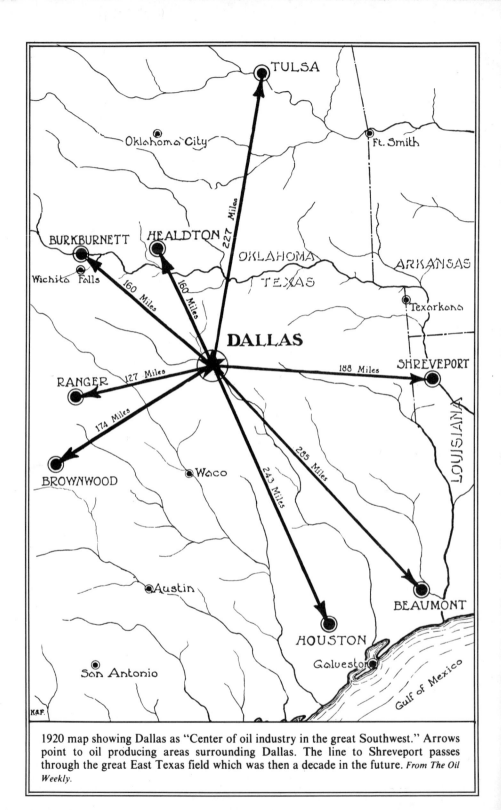

1920 map showing Dallas as "Center of oil industry in the great Southwest." Arrows point to oil producing areas surrounding Dallas. The line to Shreveport passes through the great East Texas field which was then a decade in the future. *From The Oil Weekly.*